1st Workshop on NLP for COVID-19 at ACL 2020

Online
5 – 10 July 2020

ISBN: 978-1-7138-2048-2

Introduction to the 1st Workshop on Natural Language Processing for COVID-19 at ACL 2020

Karin Verspoor, Kevin Bretonnel Cohen, Mark Dredze, Emilio Ferrara,
Jonathan May, Robert Munro, Cecile Paris, Byron Wallace

The unprecedented global pandemic related to the spread of the coronavirus SARS-COV-2 and the associated outbreak of the infection dubbed COVID-19 has had dramatic impacts worldwide during 2020. Scientists around the globe have responded to the pandemic, hoping to make some contribution to understanding, tracking, modeling, and/or responding.

The ACL community can play a unique role in supporting research to combat COVID-19. Valuable insights and critical information may be contained in vast quantities of unstructured text and speech data. Thousands of previously published research articles (and those being published on a daily basis) on coronavirus may shape our understanding of the virus or support best practice clinical management of the disease. Analysis of millions of social media posts may help us understand how the public at large is responding to the outbreak. Identifying spreading misinformation can be critical to public health messaging. Automatic identification and organization of helpful information collected from the web might aid public response.

The impetus behind organizing this "emergency" workshop was to highlight the myriad ways in which Natural Language Processing (NLP) could be used to respond to the COVID-19 pandemic, and the ACL community rose to the challenge, supported by resources such as the CORD-19 dataset from the Allen Institute for AI which was used for a Kaggle challenge[1]. We are pleased to have one of the first papers introducing this important data set amongst our accepted papers [1].

We announced the workshop on April 03, 2020 and immediately created an OpenReview site open for submissions, following an open rolling review process in which we would review as papers were submitted. We opted for single-blind reviewing, so that papers would be visible online from submission, and reviewing could proceed in an open manner. Public commentary was also enabled on the submissions, to allow for ongoing discussion of the submitted work. We received our first paper on April 09 [11], indicating just how ready the NLP community was to respond.

In all, we received 75 submissions; 50 of these arrived in the final two days before submission closed, 9 days ahead of the workshop date. With the rush of last-minute submissions, we were overwhelmed – both by the tremendous response of the community to the call, and the daunting prospect of running a rigorous review process with barely a week's turnaround to the workshop. We made the difficult decision to defer the final 50 submissions to a different process; Part 2 of the workshop is therefore now in preparation for EMNLP2020.

Of the 25 submissions that were reviewed, 17 (68%) were selected for presentation. All of these papers are included in this Proceedings volume in some form; either as an abstract only, or as a short or long paper. The topics they address range from literature mining to social media analysis.

We invited authors of the 50 deferred papers to submit posters or videos for their work, and several took us up on the opportunity. These are linked from the workshop website at `https://www.nlpcovid19workshop.org/acl2020/posters`. We also set up virtual "poster sessions" via Zoom and announced these on social media.

These months of emergency workshop organization proved to be intense but very rewarding; we are greatly appreciative to all of the authors who submitted their work, the reviewers who helped us assess submissions in a timely manner, and the broader efforts of the ACL community to encourage and enable us to pull everything together. We are proud to showcase the tremendous collective work in this Proceedings volume.

---

[1] `https://www.kaggle.com/allen-institute-for-ai/CORD-19-research-challenge`

Program Committee

Organizers

Karin Verspoor, University of Melbourne (Australia)
Kevin Bretonnel Cohen, University of Colorado Anschutz Medical Center (USA)
Mark Dredze, Johns Hopkins University (USA)
Emilio Ferrara, University of Southern California (USA)
Jonathan May, Information Sciences Institute (USA)
Robert Munro, Machine Learning Consulting (USA)
Cecile Paris, CSIRO (Australia)
Byron Wallace, Northeastern University (USA)

Reviewers

We are grateful for the huge efforts of our reviewers, who responded to the need for rapid reviews in the unfamiliar OpenReview system, as well as engaging in more interactive discussion through several rounds of author response. Quite a number of reviewers were recruited urgently after the submission site closed on June 30, and were asked to review with a turnaround of a few days. We couldn't have reviewed even the 25 submissions considered in the available time without the strong response to our (slightly desperate) requests for assistance. So a big, "Thank you!" goes out to the following reviewers:

1. Abeed Sarker, Emory University (USA)

2. Aditya Joshi, CSIRO (Australia)

3. Alexander Spangher, University of Southern California (USA)

4. Antonio Jimeno Yepes, IBM Research (Australia)

5. Benjamin E. Nye, Northeastern University (USA)

6. Berry de Bruijn, National Research Council (Canada)

7. Bevan Koopman, CSIRO (Australia)

8. Christian Boitet, IMAG (France)

9. Daniel Santel, Cincinnati Children's Medical Center (USA)

10. Danielle L. Mowery, University of Pennsylvania (USA)

11. David Lowell, Northeastern University (USA)

12. David Martinez Iraola, IBM Research (Australia)

13. Davy Weissenbacher, University of Pennsylvania (USA)

14. Diego Molla, Macquarie University (Australia)

15. Estrid He, University of Melbourne (Australia)

16. Hadi Amiri, University of Massachusetts (Lowell)

17. Haewoon Kwak, Qatar Computing Research Institute (Qatar)

18. Helen L. Johnson, University of Colorado (USA)

19. Jari Björne, University of Turku (Finland)

20. Jay DeYoung, Northeastern University (USA)

21. Jimmy Lin, University of Waterloo (Canada)

22. Jingbo Xia, Intramural Institute of Applied Mathematics Research, HZAU (China)

23. Jonathan May, Information Sciences Institute (USA)

24. Kevin Bretonnel Cohen, University of Colorado Anschutz Medical Center (USA)

25. Kirk Roberts, University of Texas Health Science Center (USA)

26. Manirupa Das, The Ohio State University (USA)

27. Maria Liakata, Warwick University (UK)

28. Matthias Gallé, Naver Labs Europe (France)

29. Mayla Rachel Boguslav, University of Colorado Anschutz Medical Center (USA)

30. Meladel Mistica, University of Melbourne (Australia)

31. Michael Conway, University of Utah (USA)

32. Oliver Baclic, Public Health Agency of Canada (Canada)

33. Olivier Bodenreider, National Institutes of Health (USA)

34. Orin Hargraves, University of Colorado (USA)

35. Peter T. Corbett, Royal Society of Chemistry

36. Pierre Zweigenbaum, LIMSI (France)

37. Qian Hu, MITRE (USA)

38. Rezarta Islamaj, National Library of Medicine (USA)

39. Robert Leaman, National Institutes of Health (USA)

40. Sarthak Jain, Northeastern University (USA)

41. Sarvnaz Karimi, CSIRO (Australia)

42. Silvio Amir, Johns Hopkins University (USA)

43. Simon Suster, University of Melbourne (Australia)

44. Srijan Kumar, Georgia Tech (USA)

45. Timothy A Miller, Harvard University (USA)

46. Tristan Naumann, Microsoft (USA)

47. Vlada Rozova, University of Melbourne (Australia)

48. William R. Hersh, Oregon Health and Sciences University (USA)

49. Zenan Zhai, University of Melbourne (Australia)

Presentation Program

The papers selected for presentation at the workshop are listed below, in thematic groups. The type of paper included in the Proceedings is indicated by (Long), (Short), or (Abs) for Abstracts.

Pre-recorded videos of some of the papers listed below are available linked from the workshop website[2].

Literature Analysis and Retrieval

COVID Question Answering

Clinical and Mental Health

Social Media

References

[1] Lucy Lu Wang, Kyle Lo, Yoganand Chandrasekhar, Russell Reas, Jiangjiang Yang, Doug Burdick, Darrin Eide, Kathryn Funk, Yannis Katsis, Rodney Michael Kinney, Yunyao Li, Ziyang Liu, William Merrill, Paul Mooney, Dewey A. Murdick, Devvret Rishi, Jerry Sheehan, Zhihong Shen, Brandon Stilson, Alex D. Wade, Kuansan Wang, Nancy Xin Ru Wang, Christopher Wilhelm, Boya Xie, Douglas M. Raymond, Daniel S. Weld, Oren Etzioni, and Sebastian Kohlmeier. CORD-19: The COVID-19 open research dataset. In *Proceedings of the 1st Workshop on NLP for COVID-19 at ACL 2020*, volume 1, Online, July 2020. Association for Computational Linguistics.

[2] Edwin Zhang, Nikhil Gupta, Rodrigo Nogueira, Kyunghyun Cho, and Jimmy Lin. Rapidly deploying a neural search engine for the COVID-19 Open Research Dataset. In *Proceedings of the 1st Workshop on NLP for COVID-19 at ACL 2020*, volume 1, Online, July 2020. Association for Computational Linguistics.

[3] Bernal Jiménez Gutiérrez, Juncheng Zeng, Dongdong Zhang, Ping Zhang, and Yu Su. Document classification for COVID-19 literature. In *Proceedings of the 1st Workshop on NLP for COVID-19 at ACL 2020*, volume 1, Online, July 2020. Association for Computational Linguistics.

[4] Alexander Spangher, Nanyun Peng, Jonathan May, and Emilio Ferrara. Enabling low-resource transfer learning across COVID-19 corpora by combining event-extraction and co-training. In *Proceedings of the 1st Workshop on NLP for COVID-19 at ACL 2020*, volume 1, Online, July 2020. Association for Computational Linguistics.

[5] Dusan Grujicic, Gorjan Radevski, Tinne Tuytelaars, and Matthew Blaschko. Self-supervised context-aware COVID-19 document exploration through atlas grounding. In *Proceedings of the 1st Workshop on NLP for COVID-19 at ACL 2020*, volume 1, Online, July 2020. Association for Computational Linguistics.

[6] Ting-Hao Kenneth Huang, Chieh-Yang Huang, Chien-Kuang Cornelia Ding, Yen-Chia Hsu, and C Lee Giles. CODA-19: Using a non-expert crowd to annotate research aspects on 10,000+ abstracts in the COVID-19 open research dataset. In *Proceedings of the 1st Workshop on NLP for COVID-19 at ACL 2020*, volume 1, Online, July 2020. Association for Computational Linguistics.

[7] Debasmita Das, Yatin Katyal, Janu Verma, Shashank Dubey, AakashDeep Singh, Kushagra Agarwal, Sourojit Bhaduri, and RajeshKumar Ranjan. Information retrieval and extraction on COVID-19 clinical articles using graph community detection and Bio-BERT embeddings. In *Proceedings of the 1st Workshop on NLP for COVID-19 at ACL 2020*, volume 1, Online, July 2020. Association for Computational Linguistics.

[8] Jerry Wei, Chengyu Huang, Soroush Vosoughi, and Jason Wei. What are people asking about COVID-19? a question classification dataset. In *Proceedings of the 1st Workshop on NLP for COVID-19 at ACL 2020*, volume 1, Online, July 2020. Association for Computational Linguistics.

[9] Yunyao Li, Tyrone Grandison, Patricia Silveyra, Ali Douraghy, Xinyu Guan, Thomas Kieselbach, Chengkai Li, and Haiqi Zhang. Jennifer for covid-19: An NLP-powered chatbot built for the people and by the people to combat misinformation. In *Proceedings of the 1st Workshop on NLP for COVID-19 at ACL 2020*, volume 1, Online, July 2020. Association for Computational Linguistics.

[10] Alec Chapman, Kelly Peterson, Augie Turano, Tamára Box, Katherine Wallace, and Makoto Jones. A natural language processing system for national COVID-19 surveillance in the us department of veterans affairs. In *Proceedings of the 1st Workshop on NLP for COVID-19 at ACL 2020*, volume 1, Online, July 2020. Association for Computational Linguistics.

[11] Bennett Kleinberg, Isabelle van der Vegt, and Maximilian Mozes. Measuring Emotions in the COVID-19 Real World Worry Dataset. In *Proceedings of the 1st Workshop on NLP for COVID-19 at ACL 2020*, volume 1, Online, July 2020. Association for Computational Linguistics.

[12] JT Wolohan. Estimating the effect of COVID-19 on mental health: Linguistic indicators of depression during a global pandemic. In *Proceedings of the 1st Workshop on NLP for COVID-19 at ACL 2020*, volume 1, Online, July. Association for Computational Linguistics.

[13] Jai Aggarwal, Ella Rabinovich, and Suzanne Stevenson. Exploration of gender differences in COVID-19 discourse on reddit. In *Proceedings of the 1st Workshop on NLP for COVID-19 at ACL 2020*, volume 1, Online, July 2020. Association for Computational Linguistics.

[14] Anna Kruspe, Matthias Häberle, Iona Kuhn, and Xiao Xiang Zhu. Cross-language sentiment analysis of European Twitter messages during the COVID-19 pandemic. In *Proceedings of the 1st Workshop on NLP for COVID-19 at ACL 2020*, volume 1, Online, July 2020. Association for Computational Linguistics.

[15] Sharon Levy and William Yang Wang. Cross-lingual transfer learning for COVID-19 outbreak alignment. In *Proceedings of the 1st Workshop on NLP for COVID-19 at ACL 2020*, volume 1, Online, July 2020. Association for Computational Linguistics.

[16] Lama Alsudias and Paul Rayson. COVID-19 and arabic twitter: How can arab world governments and public health organizations learn from social media? In *Proceedings of the 1st Workshop on NLP for COVID-19 at ACL 2020*, volume 1, Online, July 2020. Association for Computational Linguistics.

[17] Juan Carlos Medina Serrano and Orestis Papakyriakopoulos and Simon Hegelich. NLP-based Feature Extraction for the Detection of COVID-19 Misinformation Videos on YouTube. In *Proceedings of the 1st Workshop on NLP for COVID-19 at ACL 2020*, volume 1, Online, July 2020. Association for Computational Linguistics.

CORD-19: The COVID-19 Open Research Dataset

Lucy Lu Wang[1,*]   **Kyle Lo**[1,*]   **Yoganand Chandrasekhar**[1]   **Russell Reas**[1]
Jiangjiang Yang[1]   **Douglas Burdick**[2]   **Darrin Eide**[3]   **Kathryn Funk**[4]
Yannis Katsis[2]   **Rodney Kinney**[1]   **Yunyao Li**[2]   **Ziyang Liu**[6]
William Merrill[1]   **Paul Mooney**[5]   **Dewey Murdick**[7]   **Devvret Rishi**[5]
Jerry Sheehan[4]   **Zhihong Shen**[3]   **Brandon Stilson**[1]   **Alex D. Wade**[6]
Kuansan Wang[3]   **Nancy Xin Ru Wang**[2]   **Chris Wilhelm**[1]   **Boya Xie**[3]
Douglas Raymond[1]   **Daniel S. Weld**[1,8]   **Oren Etzioni**[1]   **Sebastian Kohlmeier**[1]

[1]Allen Institute for AI   [2]IBM Research   [3]Microsoft Research
[4]National Library of Medicine   [5]Kaggle   [6]Chan Zuckerberg Initiative
[7]Georgetown University   [8]University of Washington

{lucyw, kylel}@allenai.org

Abstract

The COVID-19 Open Research Dataset (CORD-19) is a growing[1] resource of scientific papers on COVID-19 and related historical coronavirus research. CORD-19 is designed to facilitate the development of text mining and information retrieval systems over its rich collection of metadata and structured full text papers. Since its release, CORD-19 has been downloaded[2] over 200K times and has served as the basis of many COVID-19 text mining and discovery systems. In this article, we describe the mechanics of dataset construction, highlighting challenges and key design decisions, provide an overview of how CORD-19 has been used, and describe several shared tasks built around the dataset. We hope this resource will continue to bring together the computing community, biomedical experts, and policy makers in the search for effective treatments and management policies for COVID-19.

1   Introduction

On March 16, 2020, the Allen Institute for AI (AI2), in collaboration with our partners at The White House Office of Science and Technology Policy (OSTP), the National Library of Medicine (NLM), the Chan Zuckerburg Initiative (CZI), Microsoft Research, and Kaggle, coordinated by Georgetown University's Center for Security and Emerging Technology (CSET), released the first version

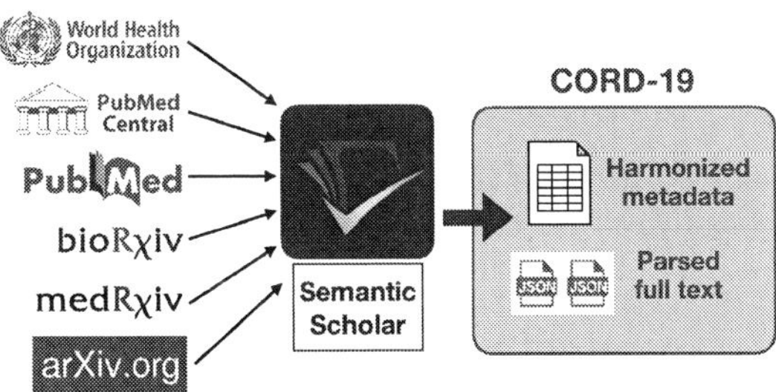

Figure 1: Papers and preprints are collected from different sources through Semantic Scholar. Released as part of CORD-19 are the harmonized and deduplicated metadata and full text JSON.

of CORD-19. This resource is a large and growing collection of publications and preprints on COVID-19 and related historical coronaviruses such as SARS and MERS. The initial release consisted of 28K papers, and the collection has grown to more than 140K papers over the subsequent weeks. Papers and preprints from several archives are collected and ingested through the Semantic Scholar literature search engine,[3] metadata are harmonized and deduplicated, and paper documents are processed through the pipeline established in Lo et al. (2020) to extract full text (more than 50% of papers in CORD-19 have full text). We commit to providing regular updates to the dataset until an end to the COVID-19 crisis is foreseeable.

CORD-19 aims to connect the machine learning community with biomedical domain experts and policy makers in the race to identify effective treatments and management policies for COVID-19. The goal is to harness these diverse and com-

---

*denotes equal contribution

[1]The dataset continues to be updated daily with papers from new sources and the latest publications. Statistics reported in this article are up-to-date as of version 2020-06-14.

[2]https://www.semanticscholar.org/cord19

[3]https://semanticscholar.org/

plementary pools of expertise to discover relevant information more quickly from the literature. Users of the dataset have leveraged AI-based techniques in information retrieval and natural language processing to extract useful information.

Responses to CORD-19 have been overwhelmingly positive, with the dataset being downloaded over 200K times in the three months since its release. The dataset has been used by clinicians and clinical researchers to conduct systematic reviews, has been leveraged by data scientists and machine learning practitioners to construct search and extraction tools, and is being used as the foundation for several successful shared tasks. We summarize research and shared tasks in Section 4.

In this article, we briefly describe:

1. The content and creation of CORD-19,
2. Design decisions and challenges around creating the dataset,
3. Research conducted on the dataset, and how shared tasks have facilitated this research, and
4. A roadmap for CORD-19 going forward.

2   Dataset

CORD-19 integrates papers and preprints from several sources (Figure 1), where a paper is defined as the base unit of published knowledge, and a preprint as an unpublished but publicly available counterpart of a paper. Throughout the rest of Section 2, we discuss papers, though the same processing steps are adopted for preprints. First, we ingest into Semantic Scholar paper metadata and documents from each source. Each paper is associated with bibliographic metadata, like title, authors, publication venue, etc, as well as unique identifiers such as a DOI, PubMed Central ID, PubMed ID, the WHO Covidence #,[4] MAG identifier (Shen et al., 2018), and others. Some papers are associated with documents, the physical artifacts containing paper content; these are the familiar PDFs, XMLs, or physical print-outs we read.

For the CORD-19 effort, we generate harmonized and deduplicated metadata as well as structured full text parses of paper documents as output. We provide full text parses in cases where we have access to the paper documents, and where the documents are available under an open access license

(e.g. Creative Commons (CC),[5] publisher-specific COVID-19 licenses,[6] or identified as open access through DOI lookup in the Unpaywall[7] database).

2.1   Sources of papers

Papers in CORD-19 are sourced from PubMed Central (PMC), PubMed, the World Health Organization's Covid-19 Database,[4] and preprint servers bioRxiv, medRxiv, and arXiv. The PMC Public Health Emergency Covid-19 Initiative[6] expanded access to COVID-19 literature by working with publishers to make coronavirus-related papers discoverable and accessible through PMC under open access license terms that allow for reuse and secondary analysis. BioRxiv and medRxiv preprints were initially provided by CZI, and are now ingested through Semantic Scholar along with all other included sources. We also work directly with publishers such as Elsevier[8] and Springer Nature,[9] to provide full text coverage of relevant papers available in their back catalog.

All papers are retrieved given the query[10]:

```
"COVID" OR "COVID-19" OR
"Coronavirus" OR "Corona virus"
OR "2019-nCoV" OR "SARS-CoV"
OR "MERS-CoV" OR "Severe Acute
Respiratory Syndrome" OR "Middle
East Respiratory Syndrome"
```

Papers that match on these keywords in their title, abstract, or body text are included in the dataset. Query expansion is performed by PMC on these search terms, affecting the subset of papers in CORD-19 retrieved from PMC.

2.2   Processing metadata

The initial collection of sourced papers suffers from duplication and incomplete or conflicting metadata. We perform the following operations to harmonize and deduplicate all metadata:

1. Cluster papers using paper identifiers
2. Select canonical metadata for each cluster
3. Filter clusters to remove unwanted entries

---

[4]https://www.who.int/emergencies/diseases/novel-coronavirus-2019/global-research-on-novel-coronavirus-2019-ncov

[5]https://creativecommons.org/

[6]https://www.ncbi.nlm.nih.gov/pmc/about/covid-19/

[7]https://unpaywall.org/

[8]https://www.elsevier.com/connect/coronavirus-information-center

[9]https://www.springernature.com/gp/researchers/campaigns/coronavirus

[10]Adapted from the Elsevier COVID-19 site[8]

Clustering papers We cluster papers if they overlap on any of the following identifiers: {*doi, pmc_id, pubmed_id, arxiv_id, who_covidence_id, mag_id*}. If two papers from different sources have an identifier in common and no other identifier conflicts between them, we assign them to the same cluster. Each cluster is assigned a unique identifier **CORD_UID**, which persists between dataset releases. No existing identifier, such as DOI or PMC ID, is sufficient as the primary CORD-19 identifier. Some papers in PMC do not have DOIs; some papers from the WHO, publishers, or preprint servers like arXiv do not have PMC IDs or DOIs.

Occasionally, conflicts occur. For example, a paper c with $(doi, pmc\_id, pubmed\_id)$ identifiers $(x, null, z')$ might share identifier x with a cluster of papers $\{a, b\}$ that has identifiers (x, y, z), but has a conflict $z' \neq z$. In this case, we choose to create a new cluster $\{c\}$, containing only paper c.[11]

Selecting canonical metadata Among each cluster, the canonical entry is selected to prioritize the availability of document files and the most permissive license. For example, between two papers with PDFs, one available under a CC license and one under a more restrictive COVID-19-specific copyright license, we select the CC-licensed paper entry as canonical. If any metadata in the canonical entry are missing, values from other members of the cluster are promoted to fill in the blanks.

Cluster filtering Some entries harvested from sources are not papers, and instead correspond to materials like tables of contents, indices, or informational documents. These entries are identified in an ad hoc manner and removed from the dataset.

2.3 Processing full text

Most papers are associated with one or more PDFs.[12] To extract full text and bibliographies from each PDF, we use the PDF parsing pipeline created for the S2ORC dataset (Lo et al., 2020).[13] In (Lo et al., 2020), we introduce the S2ORC JSON format for representing scientific paper full text,

which is used as the target output for paper full text in CORD-19. The pipeline involves:

1. Parse all PDFs to TEI XML files using GRO-BID[15] (Lopez, 2009)
2. Parse all TEI XML files to S2ORC JSON
3. Postprocess to clean up links between inline citations and bibliography entries.

We additionally parse JATS XML[16] files available for PMC papers using a custom parser, generating the same target S2ORC JSON format.

This creates two sets of full text JSON parses associated with the papers in the collection, one set originating from PDFs (available from more sources), and one set originating from JATS XML (available only for PMC papers). Each PDF parse has an associated SHA, the 40-digit SHA-1 of the associated PDF file, while each XML parse is named using its associated PMC ID. Around 48% of CORD-19 papers have an associated PDF parse, and around 37% have an XML parse, with the latter nearly a subset of the former. Most PDFs (>90%) are successfully parsed. Around 2.6% of CORD-19 papers are associated with multiple PDF SHA, due to a combination of paper clustering and the existence of supplementary PDF files.

2.4 Table parsing

Since the May 12, 2020 release of CORD-19, we also release selected HTML table parses. Tables contain important numeric and descriptive information such as sample sizes and results, which are the targets of many information extraction systems. A separate PDF table processing pipeline is used, consisting of table extraction and table understanding. *Table extraction* is based on the Smart Document Understanding (SDU) capability included in IBM Watson Discovery.[17] SDU converts a given PDF document from its native binary representation into a text-based representation like HTML which includes both identified document structures (e.g., tables, section headings, lists) and formatting information (e.g. positions for extracted text). *Table understanding* (also part of Watson Discovery) then annotates the extracted tables with additional semantic information, such as column and row headers and table captions. We leverage the Global Table Extractor (GTE) (Zheng et al.,

---

[11]This is a conservative clustering policy in which any metadata conflict prohibits clustering. An alternative policy would be to cluster if any identifier matches, under which a, b, and c would form one cluster with identifiers $(x, y, [z, z'])$.

[12]PMC papers can have multiple associated PDFs per paper, separating the main text from supplementary materials.

[13]One major difference in full text parsing for CORD-19 is that we do not use ScienceParse,[14] as we always derive this metadata from the sources directly.

[14]https://github.com/allenai/science-parse

[15]https://github.com/kermitt2/grobid

[16]https://jats.nlm.nih.gov/

[17]https://www.ibm.com/cloud/watson-discovery

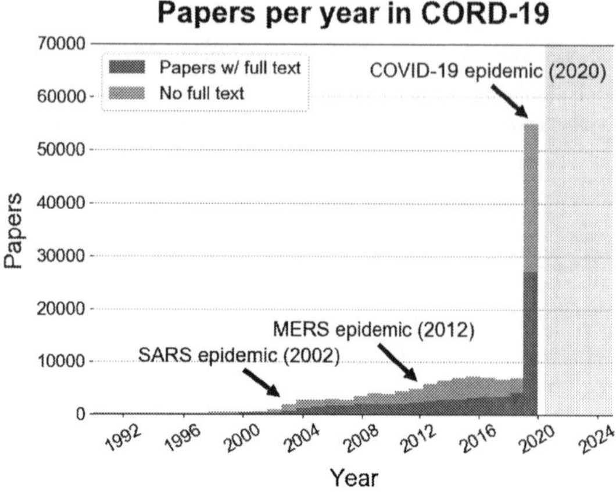

Figure 2: The distribution of papers per year in CORD-19. A spike in publications occurs in 2020 in response to COVID-19.

Subfield	Count	% of corpus
Virology	29567	25.5%
Immunology	15954	13.8%
Surgery	15667	13.5%
Internal medicine	12045	10.4%
Intensive care medicine	10624	9.2%
Molecular biology	7268	6.3%
Pathology	6611	5.7%
Genetics	5231	4.5%
Other	12997	11.2%

Table 1: MAG subfield of study for CORD-19 papers.

2020), which uses a specialized object detection and clustering technique to extract table bounding boxes and structures.

All PDFs are processed through this table extraction and understanding pipeline. If the Jaccard similarity of the table captions from the table parses and CORD-19 parses is above 0.9, we insert the HTML of the matched table into the full text JSON. We extract 188K tables from 54K documents, of which 33K tables are successfully matched to tables in 19K (around 25%) full text documents in CORD-19. Based on preliminary error analysis, we find that match failures are primarily due to caption mismatches between the two parse schemes. Thus, we plan to explore alternate matching functions, potentially leveraging table content and document location as additional features. See Appendix A for example table parses.

2.5 Dataset contents

CORD-19 has grown rapidly, now consisting of over 140K papers with over 72K full texts. Over 47K papers and 7K preprints on COVID-19 and coronaviruses have been released since the start of 2020, comprising nearly 40% of papers in the dataset.

Classification of CORD-19 papers to Microsoft Academic Graph (MAG) (Wang et al., 2019, 2020) fields of study (Shen et al., 2018) indicate that the dataset consists predominantly of papers in Medicine (55%), Biology (31%), and Chemistry (3%), which together constitute almost 90% of the corpus.[18] A breakdown of the most common MAG

subfields (L1 fields of study) represented in CORD-19 is given in Table 1.

Figure 2 shows the distribution of CORD-19 papers by date of publication. Coronavirus publications increased during and following the SARS and MERS epidemics, but the number of papers published in the early months of 2020 exploded in response to the COVID-19 epidemic. Using author affiliations in MAG, we identify the countries from which the research in CORD-19 is conducted. Large proportions of CORD-19 papers are associated with institutions based in the Americas (around 48K papers), Europe (over 35K papers), and Asia (over 30K papers).

3 Design decision & challenges

A number of challenges come into play in the creation of CORD-19. We summarize the primary design requirements of the dataset, along with challenges implicit within each requirement:

Up-to-date Hundreds of new publications on COVID-19 are released every day, and a dataset like CORD-19 can quickly become irrelevant without regular updates. CORD-19 has been updated daily since May 26. A processing pipeline that produces consistent results day to day is vital to maintaining a changing dataset. That is, the metadata and full text parsing results must be reproducible, identifiers must be persistent between releases, and changes or new features should ideally be compatible with previous versions of the dataset.

Handles data from multiple sources Papers from different sources must be integrated and harmonized. Each source has its own metadata format, which must be converted to the CORD-19 format, while addressing any missing or extraneous fields. The processing pipeline must also be flexible to adding new sources.

---

[18]MAG identifier mappings are provided as a supplement on the CORD-19 landing page.

Clean canonical metadata Because of the diversity of paper sources, duplication is unavoidable. Once paper metadata from each source is cleaned and organized into CORD-19 format, we apply the deduplication logic described in Section 2.2 to identify similar paper entries from different sources. We apply a conservative clustering algorithm, combining papers only when they have shared identifiers but no conflicts between any particular class of identifiers. We justify this because it is less harmful to retain a few duplicate papers than to remove a document that is potentially unique and useful.

Machine readable full text To provide accessible and canonical structured full text, we parse content from PDFs and associated paper documents. The full text is represented in S2ORC JSON format (Lo et al., 2020), a schema designed to preserve most relevant paper structures such as paragraph breaks, section headers, inline references, and citations. S2ORC JSON is simple to use for many NLP tasks, where character-level indices are often employed for annotation of relevant entities or spans. The text and annotation representations in S2ORC share similarities with BioC (Comeau et al., 2019), a JSON schema introduced by the BioCreative community for shareable annotations, with both formats leveraging the flexibility of character-based span annotations. However, S2ORC JSON also provides a schema for representing other components of a paper, such as its metadata fields, bibliography entries, and reference objects for figures, tables, and equations. We leverage this flexible and somewhat complete representation of S2ORC JSON for CORD-19. We recognize that converting between PDF or XML to JSON is lossy. However, the benefits of a standard structured format, and the ability to reuse and share annotations made on top of that format have been critical to the success of CORD-19.

Observes copyright restrictions Papers in CORD-19 and academic papers more broadly are made available under a variety of copyright licenses. These licenses can restrict or limit the abilities of organizations such as AI2 from redistributing their content freely. Although much of the COVID-19 literature has been made open access by publishers, the provisions on these open access licenses differ greatly across papers. Additionally, many open access licenses grant the ability to read, or "consume" the paper, but may be restrictive in

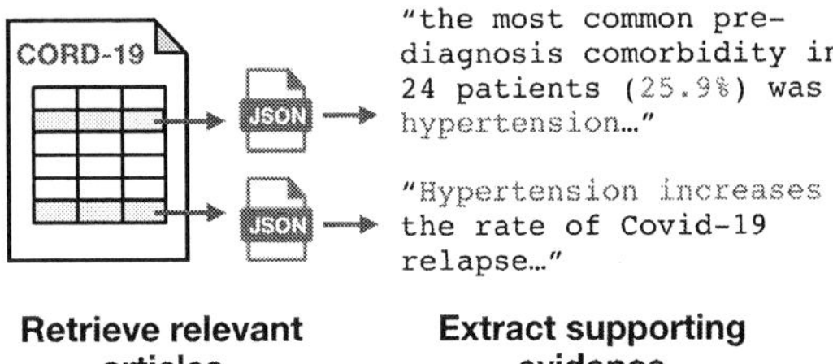

Figure 3: An example information retrieval and extraction system using CORD-19: Given an input query, the system identifies relevant papers (yellow highlighted rows) and extracts text snippets from the full text JSONs as supporting evidence.

other ways, for example, by not allowing republication of a paper or its redistribution for commercial purposes. The curator of a dataset like CORD-19 must pass on best-to-our-knowledge licensing information to the end user.

4   Research directions

We provide a survey of various ways researchers have made use of CORD-19. We organize these into four categories: *(i)* direct usage by clinicians and clinical researchers (§4.1), *(ii)* tools and systems to assist clinicians (§4.2), *(iii)* research to support further text mining and NLP research (§4.3), and *(iv)* shared tasks and competitions (§4.4).

4.1   Usage by clinical researchers

CORD-19 has been used by medical experts as a paper collection for conducting systematic reviews. These reviews address questions about COVID-19 include infection and mortality rates in different demographics (Han et al., 2020), symptoms of the disease (Parasa et al., 2020), identifying suitable drugs for repurposing (Sadegh et al., 2020), management policies (Yaacoub et al., 2020), and interactions with other diseases (Crisan-Dabija et al., 2020; Popa et al., 2020).

4.2   Tools for clinicians

Challenges for clinicians and clinical researchers during the current epidemic include *(i)* keeping up to to date with recent papers about COVID-19, *(ii)* identifying useful papers from historical coronavirus literature, *(iii)* extracting useful information from the literature, and *(iv)* synthesizing knowledge from the literature. To facilitate solutions to these challenges, dozens of tools and systems over CORD-19 have already been developed. Most combine elements of text-based information retrieval and extraction, as illustrated in Figure 3. We have compiled a list of these efforts on the CORD-19 public GitHub repository[19] and highlight some systems in Table 2.[20]

4.3 Text mining and NLP research

The following is a summary of resources released by the NLP community on top of CORD-19 to support other research activities.

Information extraction To support extractive systems, NER and entity linking of biomedical entities can be useful. NER and linking can be performed using NLP toolkits like ScispaCy (Neumann et al., 2019) or language models like BioBERT-base (Lee et al., 2019) and SciBERT-base (Beltagy et al., 2019) finetuned on biomedical NER datasets. Wang et al. (2020) augments CORD-19 full text with entity mentions predicted from several techniques, including weak supervision using the NLM's Unified Medical Language System (UMLS) Metathesaurus (Bodenreider, 2004).

Text classification Some efforts focus on extracting sentences or passages of interest. For example, Liang and Xie (2020) uses BERT (Devlin et al., 2019) to extract sentences from CORD-19 that contain COVID-19-related radiological findings.

Pretrained model weights BioBERT and SciBERT have been popular pretrained LMs for COVID-19-related tasks. DeepSet has released a BERT-base model pretrained on CORD-19.[21] SPECTER (Cohan et al., 2020) paper embeddings computed using paper titles and abstracts are being released with each CORD-19 update. SeVeN relation embeddings (Espinosa-Anke and Schockaert, 2018) between word pairs have also been made available for CORD-19.[22]

Knowledge graphs The Covid Graph project[23] releases a COVID-19 knowledge graph built from mining several public data sources, including

CORD-19, and is perhaps the largest current initiative in this space. Ahamed and Samad (2020) rely on entity co-occurrences in CORD-19 to construct a graph that enables centrality-based ranking of drugs, pathogens, and biomolecules.

4.4 Competitions and Shared Tasks

The adoption of CORD-19 and the proliferation of text mining and NLP systems built on top of the dataset are supported by several COVID-19-related competitions and shared tasks.

4.4.1 Kaggle

Kaggle hosts the CORD-19 Research Challenge,[24] a text-mining challenge that tasks participants with extracting answers to key scientific questions about COVID-19 from the papers in the CORD-19 dataset. Round 1 was initiated with a set of open-ended questions, e.g., *What is known about transmission, incubation, and environmental stability?* and *What do we know about* COVID-19 *risk factors?*

More than 500 teams participated in Round 1 of the Kaggle competition. Feedback from medical experts during Round 1 identified that the most useful contributions took the form of article summary tables. Round 2 subsequently focused on this task of table completion, and resulted in 100 additional submissions. A unique tabular schema is defined for each question, and answers are collected from across different automated extractions. For example, extractions for risk factors should include disease severity and fatality metrics, while extractions for incubation should include time ranges. Sufficient knowledge of COVID-19 is necessary to define these schema, to understand which fields are important to include (and exclude), and also to perform error-checking and manual curation.

4.4.2 TREC

The TREC-COVID[25] shared task (Roberts et al., 2020; Voorhees et al., 2020) assesses systems on their ability to rank papers in CORD-19 based on their relevance to COVID-19-related topics. Topics are sourced from MedlinePlus searches, Twitter conversations, library searches at OHSU, as well as from direct conversations with researchers, reflecting actual queries made by the community. To emulate real-world surge in publications and rapidly-

---

[19]https://github.com/allenai/cord19

[20]There are many Search and QA systems to survey. We have chosen to highlight the systems that were made publicly-available within a few weeks of the CORD-19 initial release.

[21]https://huggingface.co/deepset/covid_bert_base

[22]https://github.com/luisespinosaanke/cord-19-seven

[23]https://covidgraph.org/

[24]https://www.kaggle.com/allen-institute-for-ai/CORD-19-research-challenge

[25]https://ir.nist.gov/covidSubmit/index.html

Task	Project	Link	Description
Search and discovery	NEURAL COVIDEX	https://covidex.ai/	Uses a T5-base (Raffel et al., 2019) unsupervised reranker on BM25 (Jones et al., 2000)
	COVIDSCHOLAR	https://covidscholar.org/	Adapts Weston et al. (2019) system for entity-centric queries
	KDCOVID	http://kdcovid.nl/about.html	Uses BioSentVec (Chen et al., 2019) similarity to identify relevant sentences
	SPIKE-CORD	https://spike.covid-19.apps.allenai.org	Enables users to define "regular expression"-like queries to directly search over full text
Question answering	COVIDASK	https://covidask.korea.ac.kr/	Adapts Seo et al. (2019) using BioASQ challenge (Task B) dataset (Tsatsaronis et al., 2015)
	AUEB	http://cslab241.cs.aueb.gr:5000/	Adapts McDonald et al. (2018) using Tsatsaronis et al. (2015)
Summarization	Vespa	https://cord19.vespa.ai/	Generates summaries of paper abstracts using T5 (Raffel et al., 2019)
Recommendation	Vespa	https://cord19.vespa.ai/	Recommends "similar papers" using Sentence-BERT (Reimers and Gurevych, 2019) and SPECTER embeddings (Cohan et al., 2020)
Entailment	COVID papers browser	https://github.com/gsarti/covid-papers-browser	Similar to KDCOVID, but uses embeddings from BERT models trained on NLI datasets
Claim verification	SciFact	https://scifact.apps.allenai.org	Uses RoBERTa-large (Liu et al., 2019) to find Support/Refute evidence for COVID-19 claims
Assistive lit. review	ASReview	https://github.com/asreview/asreview-covid19	Active learning system with a CORD-19 plugin for identifying papers for literature reviews
Augmented reading	Sinequa	https://covidsearch.sinequa.com/app/covid-search/	In-browser paper reader with entity highlighting on PDFs
Visualization	SciSight	https://scisight.apps.allenai.org	Network visualizations for browsing research groups working on COVID-19

Table 2: Publicly-available tools and systems for medical experts using CORD-19.

changing information needs, the shared task is organized in multiple rounds. Each round uses a specific version of CORD-19, has newly added topics, and gives participants one week to submit per-topic document rankings for judgment. Round 1 topics included more general questions such as *What is the origin of COVID-19?* and *What are the initial symptoms of COVID-19?* while Round 3 topics have become more focused, e.g., *What are the observed mutations in the SARS-CoV-2 genome?* and *What are the longer-term complications of those who recover from COVID-19?* Around 60 medical domain experts, including indexers from NLM and medical students from OHSU and UTHealth, are involved in providing gold rankings for evaluation. TREC-COVID opened using the April 1st CORD-19 version and received submissions from over 55 participating teams.

5 Discussion

Several hundred new papers on COVID-19 are now being published every day. Automated methods are needed to analyze and synthesize information over this large quantity of content. The computing community has risen to the occasion, but it is clear that there is a critical need for better infrastructure to incorporate human judgments in the loop. Extractions need expert vetting, and search engines and systems must be designed to serve users.

Successful engagement and usage of CORD-19 speaks to our ability to bridge computing and biomedical communities over a common, global cause. From early results of the Kaggle challenge, we have learned which formats are conducive to collaboration, and which questions are the most urgent to answer. However, there is significant work that remains for determining *(i)* which methods are best to assist textual discovery over the literature, *(ii)* how best to involve expert curators in the pipeline, and *(iii)* which extracted results convert to successful COVID-19 treatments and management policies. Shared tasks and challenges, as well as continued analysis and synthesis of feedback will hopefully provide answers to these outstanding questions.

Since the initial release of CORD-19, we have implemented several new features based on community feedback, such as the inclusion of unique identifiers for papers, table parses, more sources, and daily updates. Most substantial outlying features requests have been implemented or addressed at this time. We will continue to update the dataset with more sources of papers and newly published literature as resources permit.

5.1 Limitations

Though we aim to be comprehensive, CORD-19 does not cover many relevant scientific documents on COVID-19. We have restricted ourselves to research papers and preprints, and do not incorporate other types of documents, such as technical reports, white papers, informational publications by governmental bodies, and more. Including these documents is outside the current scope of CORD-19, but we encourage other groups to curate and publish such datasets.

Within the scope of scientific papers, CORD-19 is also incomplete, though we continue to prioritize the addition of new sources. This has motivated the creation of other corpora supporting COVID-19 NLP, such as LitCovid (Chen et al., 2020), which provide complementary materials to CORD-19 derived from PubMed. Though we have since added PubMed as a source of papers in CORD-19, there are other domains such as the social sciences that are not currently represented, and we hope to incorporate these works as part of future work.

We also note the shortage of foreign language papers in CORD-19, especially Chinese language papers produced during the early stages of the epidemic. These papers may be useful to many researchers, and we are working with collaborators to provide them as supplementary data. However, challenges in both sourcing and licensing these papers for re-publication are additional hurdles.

5.2 Call to action

Though the full text of many scientific papers are available to researchers through CORD-19, a number of challenges prevent easy application of NLP and text mining techniques to these papers. First, the primary distribution format of scientific papers – PDF – is not amenable to text processing. The PDF file format is designed to share electronic documents rendered faithfully for reading and printing, and mixes visual with semantic information. Significant effort is needed to coerce PDF into a format more amenable to text mining, such as JATS XML,[26] BioC (Comeau et al., 2019), or S2ORC JSON (Lo et al., 2020), which is used in CORD-19. Though there is substantial work in this domain, we can still benefit from better PDF parsing tools for scientific documents. As a complement, scientific papers should also be made available in a structured format like JSON, XML, or HTML.

Second, there is a clear need for more scientific content to be made accessible to researchers. Some publishers have made COVID-19 papers openly available during this time, but both the duration and scope of these epidemic-specific licenses are unclear. Papers describing research in related areas (e.g., on other infectious diseases, or relevant biological pathways) have also not been made open access, and are therefore unavailable in CORD-19 or otherwise. Securing release rights for papers not yet in CORD-19 but relevant for COVID-19 research is a significant portion of future work, led by the PMC COVID-19 Initiative.[6]

Lastly, there is no standard format for representing paper metadata. Existing schemas like the JATS XML NISO standard[26] or library science standards like BIBFRAME[27] or Dublin Core[28] have been adopted to represent paper metadata. However, these standards can be too coarse-grained to capture all necessary paper metadata elements, or may lack a strict schema, causing representations to vary greatly across publishers who use them. To improve metadata coherence across sources, the community must define and agree upon an appropriate standard of representation.

Summary

This project offers a paradigm of how the community can use machine learning to advance scientific research. By allowing computational access to the papers in CORD-19, we increase our ability to perform discovery over these texts. We hope the dataset and projects built on the dataset will serve as a template for future work in this area. We also believe there are substantial improvements that can be made in the ways we publish, share, and work with scientific papers. We offer a few suggestions that could dramatically increase community productivity, reduce redundant effort, and result in better discovery and understanding of the scientific literature.

---

[26]https://www.niso.org/publications/z3996-2019-jats
[27]https://www.loc.gov/bibframe/
[28]https://www.dublincore.org/specifications/dublin-core/dces/

Through CORD-19, we have learned the importance of bringing together different communities around the same scientific cause. It is clearer than ever that automated text analysis is not the solution, but rather one tool among many that can be directed to combat the COVID-19 epidemic. Crucially, the systems and tools we build must be designed to serve a use case, whether that's improving information retrieval for clinicians and medical professionals, summarizing the conclusions of the latest observational research or clinical trials, or converting these learnings to a format that is easily digestible by healthcare consumers.

Acknowledgments

This work was supported in part by NSF Convergence Accelerator award 1936940, ONR grant N00014-18-1-2193, and the University of Washington WRF/Cable Professorship.

We thank The White House Office of Science and Technology Policy, the National Library of Medicine at the National Institutes of Health, Microsoft Research, Chan Zuckerberg Initiative, and Georgetown University's Center for Security and Emerging Technology for co-organizing the CORD-19 initiative. We thank Michael Kratsios, the Chief Technology Officer of the United States, and The White House Office of Science and Technology Policy for providing the initial seed set of questions for the Kaggle CORD-19 research challenge.

We thank Kaggle for coordinating the CORD-19 research challenge. In particular, we acknowledge Anthony Goldbloom for providing feedback on CORD-19 and for involving us in discussions around the Kaggle literature review tables project. We thank the National Institute of Standards and Technology (NIST), National Library of Medicine (NLM), Oregon Health and Science University (OHSU), and University of Texas Health Science Center at Houston (UTHealth) for co-organizing the TREC-COVID shared task. In particular, we thank our co-organizers – Steven Bedrick (OHSU), Aaron Cohen (OHSU), Dina Demner-Fushman (NLM), William Hersh (OHSU), Kirk Roberts (UTHealth), Ian Soboroff (NIST), and Ellen Voorhees (NIST) – for feedback on the design of CORD-19.

We acknowledge our partners at Elsevier and Springer Nature for providing additional full text coverage of papers included in the corpus.

We thank Bryan Newbold from the Internet Archive for providing feedback on data quality and helpful comments on early drafts of the manuscript.

We thank Rok Jun Lee, Hrishikesh Sathe, Dhaval Sonawane and Sudarshan Thitte from IBM Watson AI for their help in table parsing.

We also acknowledge and thank our collaborators from AI2: Paul Sayre and Sam Skjonsberg for providing front-end support for CORD-19 and TREC-COVID, Michael Schmitz for setting up the CORD-19 Discourse community forums, Adriana Dunn for creating webpage content and marketing, Linda Wagner for collecting community feedback, Jonathan Borchardt, Doug Downey, Tom Hope, Daniel King, and Gabriel Stanovsky for contributing supplemental data to the CORD-19 effort, Alex Schokking for his work on the Semantic Scholar COVID-19 Research Feed, Darrell Plessas for technical support, and Carissa Schoenick for help with public relations.

References

Sabber Ahamed and Manar D. Samad. 2020. Information mining for covid-19 research from a large volume of scientific literature. *ArXiv*, abs/2004.02085.

Iz Beltagy, Kyle Lo, and Arman Cohan. 2019. SciBERT: A pretrained language model for scientific text. In *Proceedings of the 2019 Conference on Empirical Methods in Natural Language Processing and the 9th International Joint Conference on Natural Language Processing (EMNLP-IJCNLP)*, pages 3615–3620, Hong Kong, China. Association for Computational Linguistics.

Olivier Bodenreider. 2004. The unified medical language system (umls): integrating biomedical terminology. *Nucleic acids research*, 32 Database issue:D267–70.

Q. Chen, Y. Peng, and Z. Lu. 2019. Biosentvec: creating sentence embeddings for biomedical texts. In *2019 IEEE International Conference on Healthcare Informatics (ICHI)*, pages 1–5.

Qingyu Chen, Alexis Allot, and Zhiyong Lu. 2020. Keep up with the latest coronavirus research. *Nature*, 579:193 – 193.

Arman Cohan, Sergey Feldman, Iz Beltagy, Doug Downey, and Daniel S. Weld. 2020. Specter: Document-level representation learning using citation-informed transformers. In *ACL*.

Donald C. Comeau, Chih-Hsuan Wei, Rezarta Islamaj Dogan, and Zhiyong Lu. 2019. Pmc text mining subset in bioc: about three million full-text articles and growing. *Bioinformatics*.

Radu Crisan-Dabija, Cristina Grigorescu, Cristina Alice Pavel, Bogdan Artene, Iolanda Valentina Popa, Andrei Cernomaz, and Alexandru Burlacu. 2020. Tuberculosis and covid-19 in 2020: lessons from the past viral outbreaks and possible future outcomes. *medRxiv*.

Jacob Devlin, Ming-Wei Chang, Kenton Lee, and Kristina Toutanova. 2019. BERT: Pre-training of deep bidirectional transformers for language understanding. In *Proceedings of the 2019 Conference of the North American Chapter of the Association for Computational Linguistics: Human Language Technologies, Volume 1 (Long and Short Papers)*, pages 4171–4186, Minneapolis, Minnesota. Association for Computational Linguistics.

Luis Espinosa-Anke and Steven Schockaert. 2018. SeVeN: Augmenting word embeddings with unsupervised relation vectors. In *Proceedings of the 27th International Conference on Computational Linguistics*, pages 2653–2665, Santa Fe, New Mexico, USA. Association for Computational Linguistics.

M. Fathi, Khatoon Vakili, Fatemeh Sayehmiri, Abdolrahman Mohamadkhani, M. Hajiesmaeili, Mostafa Rezaei-Tavirani, and Owrang Eilami. 2020. Prognostic value of comormidity for severity of covid-19: A systematic review and meta-analysis study. In *medRxiv*.

Yang Han, Victor O.K. Li, Jacqueline C.K. Lam, Peiyang Guo, Ruiqiao Bai, and Wilton W.T. Fok. 2020. Who is more susceptible to covid-19 infection and mortality in the states? *medRxiv*.

Torsten Hothorn, Marie-Charlotte Bopp, H. F. Guenthard, Olivia Keiser, Michel Roelens, Caroline E Weibull, and Michael J Crowther. 2020. Relative coronavirus disease 2019 mortality: A swiss population-based study. In *medRxiv*.

Karen Spärck Jones, Steve Walker, and Stephen E. Robertson. 2000. A probabilistic model of information retrieval: development and comparative experiments - part 1. *Inf. Process. Manag.*, 36:779–808.

Shubhi Kaushik, Scott I. Aydin, Kim R. Derespina, Prerna Bansal, Shanna Kowalsky, Rebecca Trachtman, Jennifer K. Gillen, Michelle M. Perez, Sara H. Soshnick, Edward E. Conway, Asher Bercow, Howard S. Seiden, Robert H Pass, Henry Michael Ushay, George Ofori-Amanfo, and Shivanand S Medar. 2020. Multisystem inflammatory syndrome in children (mis-c) associated with sars-cov-2 infection: A multi-institutional study from new york city. *The Journal of Pediatrics*.

Jinhyuk Lee, Wonjin Yoon, Sungdong Kim, Donghyeon Kim, Sunkyu Kim, Chan Ho So, and Jaewoo Kang. 2019. BioBERT: a pre-trained biomedical language representation model for biomedical text mining. *Bioinformatics*.

Yuxiao Liang and Pengtao Xie. 2020. Identifying radiological findings related to covid-19 from medical literature. *ArXiv*, abs/2004.01862.

Yinhan Liu, Myle Ott, Naman Goyal, Jingfei Du, Mandar Joshi, Danqi Chen, Omer Levy, Mike Lewis, Luke Zettlemoyer, and Veselin Stoyanov. 2019. Roberta: A robustly optimized bert pretraining approach.

Kyle Lo, Lucy Lu Wang, Mark Neumann, Rodney Kinney, and Daniel S. Weld. 2020. S2ORC: The Semantic Scholar Open Research Corpus. In *Proceedings of ACL*.

Patrice Lopez. 2009. Grobid: Combining automatic bibliographic data recognition and term extraction for scholarship publications. In *ECDL*.

Luis López-Fando, Paulina Bueno, David Sánchez Carracedo, Márcio Augusto Averbeck, David Manuel Castro-Díaz, emmanuel chartier-kastler, Francisco Cruz, Roger R Dmochowski, Enrico Finazzi-Agrò, Sakineh Hajebrahimi, John Heesakkers, George R Kasyan, Tufan Tarcan, Benoît Peyronnet, Mauricio Plata, Bárbara Padilla-Fernández, Frank Van der Aa, Salvador Arlandis, and Hashim Hashim. 2020. Management of female and functional urology patients during the covid pandemic. *European Urology Focus*.

Ryan McDonald, Georgios-Ioannis Brokos, and Ion Androutsopoulos. 2018. Deep relevance ranking using enhanced document-query interactions. In *EMNLP*.

Mark Neumann, Daniel King, Iz Beltagy, and Waleed Ammar. 2019. ScispaCy: Fast and robust models for biomedical natural language processing. In *Proceedings of the 18th BioNLP Workshop and Shared Task*, pages 319–327, Florence, Italy. Association for Computational Linguistics.

Sravanthi Parasa, Madhav Desai, Viveksandeep Thoguluva Chandrasekar, Harsh Patel, Kevin Kennedy, Thomas Rösch, Marco Spadaccini, Matteo Colombo, Roberto Gabbiadini, Everson L. A. Artifon, Alessandro Repici, and Prateek Sharma. 2020. Prevalence of gastrointestinal symptoms and fecal viral shedding in patients with coronavirus disease 2019. *JAMA Network Open*, 3.

Iolanda Valentina Popa, Mircea Diculescu, Catalina Mihai, Cristina Cijevschi-Prelipcean, and Alexandru Burlacu. 2020. Covid-19 and inflammatory bowel diseases: risk assessment, shared molecular pathways and therapeutic challenges. *medRxiv*.

Colin Raffel, Noam Shazeer, Adam Roberts, Katherine Lee, Sharan Narang, Michael Matena, Yanqi Zhou, Wei Li, and Peter J. Liu. 2019. Exploring the limits of transfer learning with a unified text-to-text transformer.

Nils Reimers and Iryna Gurevych. 2019. Sentence-BERT: Sentence embeddings using Siamese BERT-networks. In *Proceedings of the 2019 Conference on Empirical Methods in Natural Language Processing and the 9th International Joint Conference on Natural Language Processing (EMNLP-IJCNLP)*, pages 3982–3992, Hong Kong, China. Association for Computational Linguistics.

Kirk Roberts, Tasmeer Alam, Steven Bedrick, Dina Demner-Fushman, Kyle Lo, Ian Soboroff, Ellen Voorhees, Lucy Lu Wang, and William R Hersh. 2020. TREC-COVID: Rationale and Structure of an Information Retrieval Shared Task for COVID-19. *Journal of the American Medical Informatics Association*. Ocaa091.

Sepideh Sadegh, Julian Matschinske, David B. Blumenthal, Gihanna Galindez, Tim Kacprowski, Markus List, Reza Nasirigerdeh, Mhaned Oubounyt, Andreas Pichlmair, Tim Daniel Rose, Marisol Salgado-Albarrán, Julian Späth, Alexey Stukalov, Nina K. Wenke, Kevin Yuan, Josch K. Pauling, and Jan Baumbach. 2020. Exploring the sars-cov-2 virus-host-drug interactome for drug repurposing.

Minjoon Seo, Jinhyuk Lee, Tom Kwiatkowski, Ankur Parikh, Ali Farhadi, and Hannaneh Hajishirzi. 2019. Real-time open-domain question answering with dense-sparse phrase index. In *Proceedings of the 57th Annual Meeting of the Association for Computational Linguistics*, pages 4430–4441, Florence, Italy. Association for Computational Linguistics.

Zhihong Shen, Hao Ma, and Kuansan Wang. 2018. A web-scale system for scientific knowledge exploration. In *Proceedings of ACL 2018, System Demonstrations*, pages 87–92, Melbourne, Australia. Association for Computational Linguistics.

Silvia Stringhini, Ania Wisniak, Giovanni Piumatti, Andrew S. Azman, Stephen A Lauer, Hélène Baysson, David De Ridder, Dusan Petrovic, Stephanie Schrempft, Kailing Marcus, Sabine Yerly, Isabelle Arm Vernez, Olivia Keiser, Samia Hurst, Klara M Posfay-Barbe, Didier Trono, Didier Pittet, Laurent Gétaz, François Chappuis, Isabella Eckerle, Nicolas Vuilleumier, Benjamin Meyer, Antoine Flahault, Laurent Kaiser, and Idris Guessous. 2020. Seroprevalence of anti-sars-cov-2 igg antibodies in geneva, switzerland (serocov-pop): a population-based study. *Lancet (London, England)*.

George Tsatsaronis, Georgios Balikas, Prodromos Malakasiotis, Ioannis Partalas, Matthias Zschunke, Michael R. Alvers, Dirk Weissenborn, Anastasia Krithara, Sergios Petridis, Dimitris Polychronopoulos, Yannis Almirantis, John Pavlopoulos, Nicolas Baskiotis, Patrick Gallinari, Thierry Artières, Axel-Cyrille Ngonga Ngomo, Norman Heino, Éric Gaussier, Liliana Barrio-Alvers, Michael Schroeder, Ion Androutsopoulos, and Georgios Paliouras. 2015. An overview of the bioasq large-scale biomedical semantic indexing and question answering competition. In *BMC Bioinformatics*.

Ellen Voorhees, Tasmeer Alam, Steven Bedrick, Dina Demner-Fushman, William R Hersh, Kyle Lo, Kirk Roberts, Ian Soboroff, and Lucy Lu Wang. 2020. TREC-COVID: Constructing a pandemic information retrieval test collection. *SIGIR Forum*, 54.

Kuansan Wang, Zhihong Shen, Chiyuan Huang, Chieh-Han Wu, Yuxiao Dong, and Anshul Kanakia. 2020. Microsoft academic graph: When experts are not enough. *Quantitative Science Studies*, 1(1):396–413.

Kuansan Wang, Zhihong Shen, Chiyuan Huang, Chieh-Han Wu, Darrin Eide, Yuxiao Dong, Junjie Qian, Anshul Kanakia, Alvin Chen, and Richard Rogahn. 2019. A review of microsoft academic services for science of science studies. *Frontiers in Big Data*, 2.

Xuan Wang, Xiangchen Song, Yingjun Guan, Bangzheng Li, and Jiawei Han. 2020. Comprehensive named entity recognition on cord-19 with distant or weak supervision. *ArXiv*, abs/2003.12218.

Leigh Weston, Vahe Tshitoyan, John Dagdelen, Olga Kononova, Kristin Persson, Gerbrand Ceder, and Anubhav Jain. 2019. Named Entity Recognition and Normalization Applied to Large-Scale Information Extraction from the Materials Science Literature.

Sally Yaacoub, Holger J Schünemann, Joanne Khabsa, Amena El-Harakeh, Assem M Khamis, Fatimah Chamseddine, Rayane El Khoury, Zahra Saad, Layal Hneiny, Carlos Cuello Garcia, Giovanna Elsa Ute Muti-Schünemann, Antonio Bognanni, Chen Chen, Guang Chen, Yuan Zhang, Hong Zhao, Pierre Abi Hanna, Mark Loeb, Thomas Piggott, Marge Reinap, Nesrine Rizk, Rosa Stalteri, Stephanie Duda, Karla Solo, Derek K Chu, and Elie A Akl. 2020. Safe management of bodies of deceased persons with suspected or confirmed covid-19: a rapid systematic review. *BMJ Global Health*, 5(5).

Xinyi Zheng, Doug Burdick, Lucian Popa, and Xin Ru Nancy Wang. 2020. Global table extractor (gte): A framework for joint table identification and cell structure recognition using visual context. *ArXiv*, abs/2005.00589.

A Table parsing results

There is high variance in the representation of tables across different paper PDFs. The goal of table parsing is to extract all tables from PDFs and represent them in HTML table format, along with associated titles and headings. In Table 3, we provide several example table parses, showing the high diversity of table representations across documents, the structure of resulting parses, and some common parse errors.

PDF Representation	HTML Table Parse	Source & Description

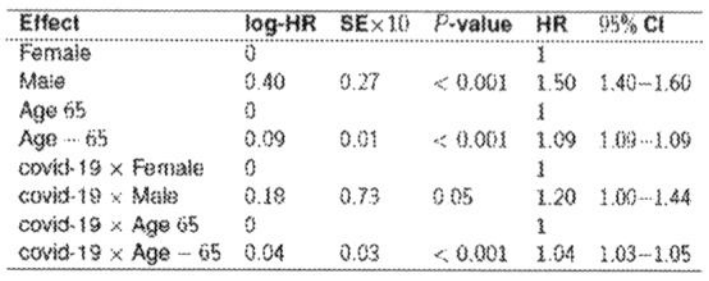

Example 1

Effect	log-HR	SE×10	P-value	HR	95% CI
Female	0				1
Male	0.40	0.27	< 0.001	1.50	1.40–1.60
Age 65	0				1
Age – 65	0.09	0.01	< 0.001	1.09	1.09–1.09
covid-19 × Female	0				1
covid-19 × Male	0.18	0.73	0.05	1.20	1.00–1.44
covid-19 × Age 65	0				1
covid-19 × Age – 65	0.04	0.03	< 0.001	1.04	1.03–1.05

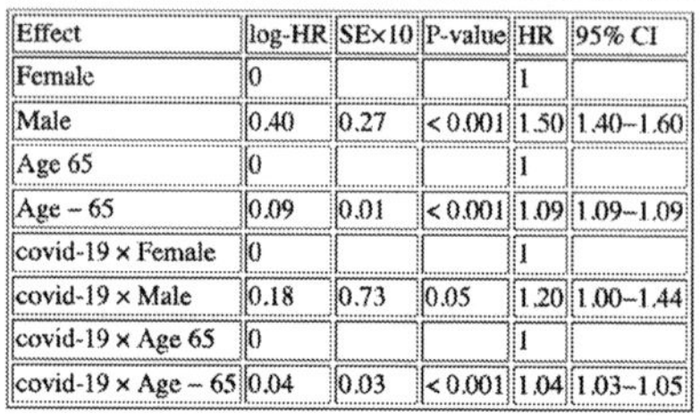

From Hothorn et al. (2020): Exact Structure; Minimal row rules

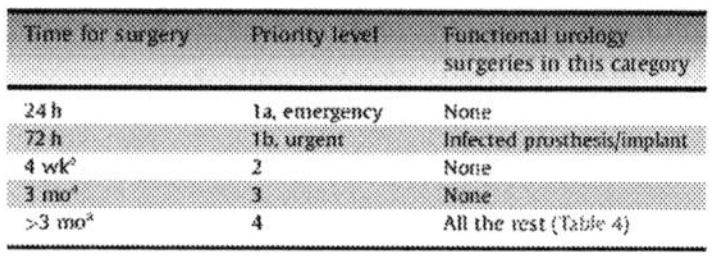

Example 2

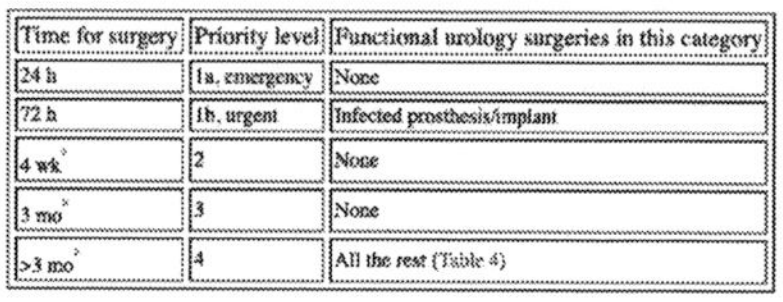

Time for surgery	Priority level	Functional urology surgeries in this category
24 h	1a, emergency	None
72 h	1b, urgent	Infected prosthesis/implant
4 wk[2]	2	None
3 mo[3]	3	None
>3 mo[4]	4	All the rest (Table 4)

From López-Fando et al. (2020): Exact Structure; Colored rows

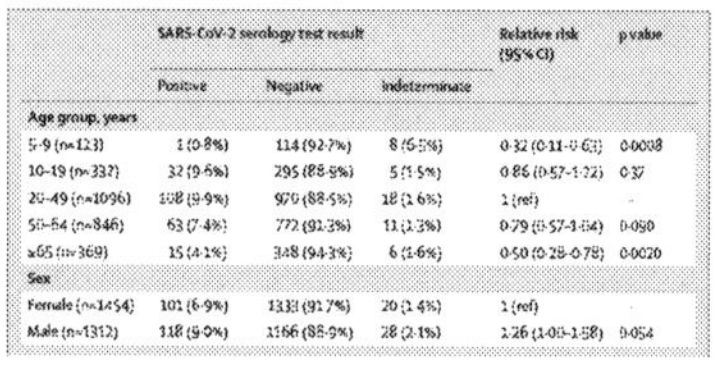

Example 3

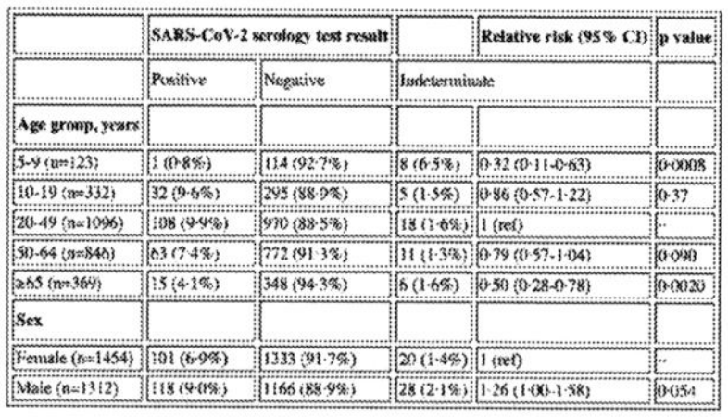

	SARS-CoV-2 serology test result			Relative risk (95% CI)	p value
	Positive	Negative	Indeterminate		
Age group, years					
5-9 (n=123)	1 (0·8%)	114 (92·7%)	8 (6·5%)	0·32 (0·11–0·63)	0·0008
10-19 (n=332)	32 (9·6%)	295 (88·9%)	5 (1·5%)	0·86 (0·57–1·22)	0·37
20-49 (n=1096)	108 (9·9%)	970 (88·5%)	18 (1·6%)	1 (ref)	..
50-64 (n=846)	63 (7·4%)	772 (91·3%)	11 (1·3%)	0·79 (0·57–1·04)	0·090
≥65 (n=369)	15 (4·1%)	348 (94·3%)	6 (1·6%)	0·50 (0·28–0·78)	0·0020
Sex					
Female (n=1454)	101 (6·9%)	1333 (91·7%)	20 (1·4%)	1 (ref)	..
Male (n=1312)	118 (9·0%)	1166 (88·9%)	28 (2·1%)	1·26 (1·00–1·58)	0·054

From Stringhini et al. (2020): Minor span errors; Partially colored background with minimal row rules

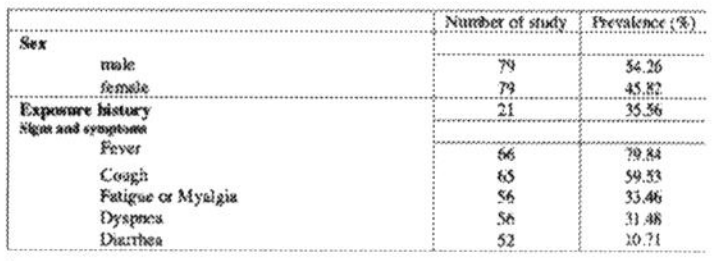

Example 4

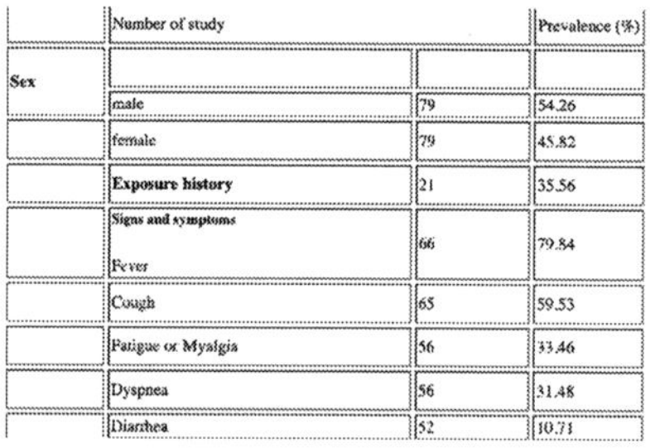

	Number of study	Prevalence (%)
Sex		
male	79	54.26
female	79	45.82
Exposure history	21	35.56
Signs and symptoms		
Fever	66	79.84
Cough	65	59.53
Fatigue or Myalgia	56	33.46
Dyspnea	56	31.48
Diarrhea	52	10.71

From Fathi et al. (2020): Overmerge and span errors; Some section headers have row rules

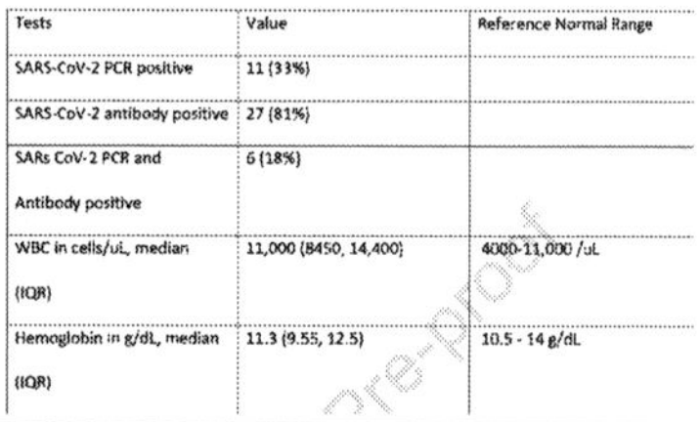

Example 5

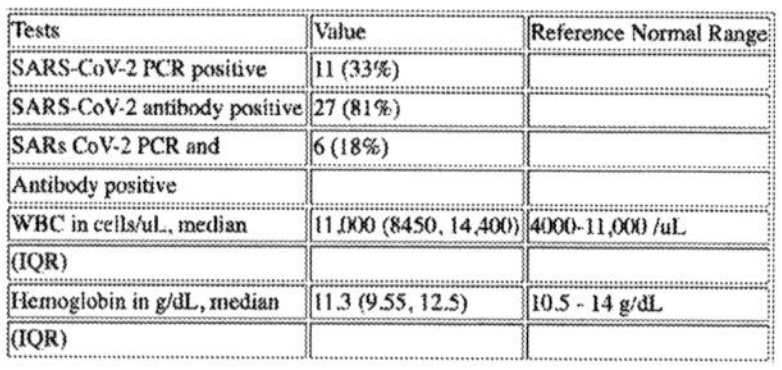

Tests	Value	Reference Normal Range
SARS-CoV-2 PCR positive	11 (33%)	
SARS-CoV-2 antibody positive	27 (81%)	
SARs CoV-2 PCR and	6 (18%)	
Antibody positive		
WBC in cells/uL, median	11,000 (8450, 14,400)	4000-11,000 /uL
(IQR)		
Hemoglobin in g/dL, median	11.3 (9.55, 12.5)	10.5 - 14 g/dL
(IQR)		

From Kaushik et al. (2020): Over-splitting errors; Full row and column rules with large vertical spacing in cells

Table 3: A sample of table parses. Though most table structure is preserved accurately, the diversity of table representations results in some errors.

Rapidly Deploying a Neural Search Engine for the COVID-19 Open Research Dataset

Edwin Zhang,[1] **Nikhil Gupta,**[1] **Rodrigo Nogueira,**[1] **Kyunghyun Cho,**[2,3,4] and **Jimmy Lin**[1]

[1] David R. Cheriton School of Computer Science, University of Waterloo
[2] Courant Institute of Mathematical Sciences, New York University
[3] Center for Data Science, New York University   [4] CIFAR Associate Fellow

This extended abstract represents an abridged version of Zhang et al. (2020a), posted on arXiv April 10, 2020 and concurrently submitted to this workshop. We have intentionally decided for this short piece to reflect the state of our work at that time. The latest updates on our project can be found in Zhang et al. (2020b).

The Neural Covidex is a search engine that exploits the latest neural ranking architectures to provide information access to the COVID-19 Open Research Dataset (CORD-19) curated by the Allen Institute for AI (Wang et al., 2020). It exists as part of a suite of tools we have developed to help domain experts tackle the ongoing global pandemic. We hope that improved information access capabilities to the scientific literature can inform evidence-based decision making and insight generation.

The first version of CORD-19 was released on March 13, 2020. Within a couple of weeks, our team was able to build, deploy, and share with the research community a number of open-source components that support information access to this corpus. These include: Extensions to our Anserini IR toolkit (Yang et al., 2018) and its Pyserini Python interface (Akkalyoncu Yilmaz et al., 2020) to support basic keyword search capabilities on the corpus; PyGaggle, a new library for neural text ranking that includes supervised ranking models based on T5 as well as unsupervised sentence highlighting models with BioBERT (Lee et al., 2020).

We have assembled these components into the Neural Covidex, available online at `covidex.ai`; see screenshot in Figure 1. This user interface was developed from scratch and is itself open source. Zhang et al. (2020a) described our initial efforts and shared lessons we learned along the way.

Although the application of BERT to text ranking is well known (Nogueira and Cho, 2019), we decided to deploy our latest research based on sequence-to-sequence models (Nogueira et al.,

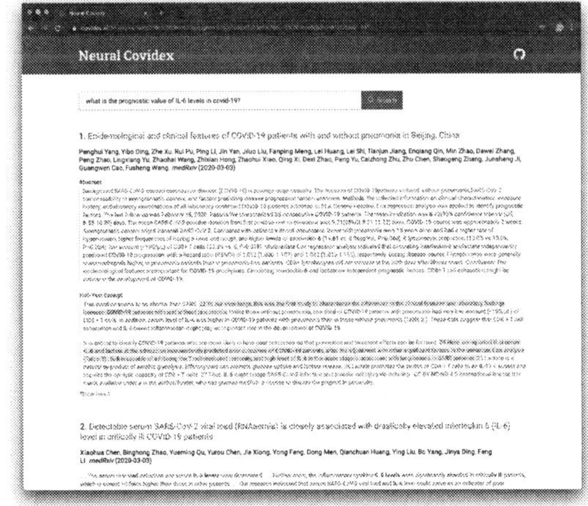

Figure 1: Screenshot of the Neural Covidex.

2020), specifically T5 (Raffel et al., 2019). This relevance classifier, which reranks BM25 results from Anserini, is fed a query q and each candidate document d in turn. The model is fine-tuned to produce either "true" or "false" depending on whether the document is relevant or not to the query. At inference time, we softmax the logits of the "true" and "false" tokens, and the resulting probability of the "true" token is used as the relevance score of d. Candidate documents are then reranked using their relevance scores. As there is no COVID-19 training data, we fine-tuned our model on the MS MARCO passage dataset (Nguyen et al., 2016), and thus our reranker operates in a zero-shot setting.

As our work pre-dated any systematic evaluation efforts by the community, at the time of submission we were unable to provide any experimental results. Since then, however, we have participated in the TREC-COVID challenge (Voorhees et al., 2020); partial results to date are reported in Zhang et al. (2020b). Nevertheless, the speed at which we were able to build and deploy the Neural Covidex is a testament to the power of open-source software and modern open-science norms.

Acknowledgments

This research was supported in part by the Canada First Research Excellence Fund, the Natural Sciences and Engineering Research Council (NSERC) of Canada, NVIDIA, eBay, Microsoft, and CIFAR. We'd like to thank Kyle Lo for helpful discussions and Colin Raffel for his assistance with T5.

References

Zeynep Akkalyoncu Yilmaz, Charles L. A. Clarke, and Jimmy Lin. 2020. A lightweight environment for learning experimental IR research practices. In *Proceedings of the 43rd Annual International ACM SIGIR Conference on Research and Development in Information Retrieval (SIGIR 2020)*, pages 2113–2116.

Jinhyuk Lee, Wonjin Yoon, Sungdong Kim, Donghyeon Kim, Sunkyu Kim, Chan Ho So, and Jaewoo Kang. 2020. BioBERT: a pretrained biomedical language representation model for biomedical text mining. *Bioinformatics*, 36(4):1234–1240.

Tri Nguyen, Mir Rosenberg, Xia Song, Jianfeng Gao, Saurabh Tiwary, Rangan Majumder, and Li Deng. 2016. MS MARCO: A human generated machine reading comprehension dataset. *arXiv:1611.09268*.

Rodrigo Nogueira and Kyunghyun Cho. 2019. Passage re-ranking with BERT. *arXiv:1901.04085*.

Rodrigo Nogueira, Zhiying Jiang, and Jimmy Lin. 2020. Document ranking with a pretrained sequence-to-sequence model. *arXiv:2003.06713*.

Colin Raffel, Noam Shazeer, Adam Roberts, Katherine Lee, Sharan Narang, Michael Matena, Yanqi Zhou, Wei Li, and Peter J. Liu. 2019. Exploring the limits of transfer learning with a unified text-to-text transformer. *arXiv:1910.10683*.

Ellen Voorhees, Tasmeer Alam, Steven Bedrick, Dina Demner-Fushman, William R. Hersh, Kyle Lo, Kirk Roberts, Ian Soboroff, and Lucy Lu Wang. 2020. TREC-COVID: Constructing a pandemic information retrieval test collection. *SIGIR Forum*, 54(1).

Lucy Lu Wang, Kyle Lo, Yoganand Chandrasekhar, Russell Reas, Jiangjiang Yang, Doug Burdick, Darrin Eide, Kathryn Funk, Yannis Katsis, Rodney Kinney, Yunyao Li, Ziyang Liu, William Merrill, Paul Mooney, Dewey Murdick, Devvret Rishi, Jerry Sheehan, Zhihong Shen, Brandon Stilson, Alex Wade, Kuansan Wang, Nancy Xin Ru Wang, Chris Wilhelm, Boya Xie, Douglas Raymond, Daniel S. Weld, Oren Etzioni, and Sebastian Kohlmeier. 2020. CORD-19: The COVID-19 Open Research Dataset. *arXiv:2004.10706*.

Peilin Yang, Hui Fang, and Jimmy Lin. 2018. Anserini: reproducible ranking baselines using Lucene. *Journal of Data and Information Quality*, 10(4):Article 16.

Edwin Zhang, Nikhil Gupta, Rodrigo Nogueira, Kyunghyun Cho, and Jimmy Lin. 2020a. Rapidly deploying a neural search engine for the COVID-19 Open Research Dataset: Preliminary thoughts and lessons learned. *arXiv:2004.05125*.

Edwin Zhang, Nikhil Gupta, Raphael Tang, Xiao Han, Ronak Pradeep, Kuang Lu, Yue Zhang, Rodrigo Nogueira, Kyunghyun Cho, Hui Fang, and Jimmy Lin. 2020b. Covidex: Neural ranking models and keyword search infrastructure for the COVID-19 Open Research Dataset. *arXiv:2007.07846*.

Document Classification for COVID-19 Literature

Bernal Jiménez Gutiérrez, Juncheng Zeng, Dongdong Zhang, Ping Zhang, Yu Su
The Ohio State University
`{jimenezgutierrez.1,zeng.671,zhang.11069,`
`zhang.10631,su.809}@osu.edu`

The global pandemic has made it more important than ever to quickly and accurately retrieve relevant scientific literature for effective consumption by researchers in a wide range of fields. We provide an analysis of several multi-label document classification models on the LitCovid dataset, a growing collection of 23,000 research papers regarding the novel 2019 coronavirus. Additionally, we test these models on a subset of the CORD-19 dataset containing 100 papers about previous epidemics we manually annotated.

Class	LitCovid	CORD-19 Set
Prevention	11,042	12
Treatment	6,897	20
Diagnosis	4,754	25
Mechanism	3,549	70
Case Report	1,914	2
Transmission	1,065	6
General	368	7
Forecasting	461	2

Table 1: Category distribution for the *LitCovid* and *CORD-19 Test Datasets*.

We find that pre-trained language models fine-tuned on this dataset outperform all other baselines and that BioBERT surpasses the others by a small margin with micro-F1 and accuracy scores of around 86% and 75% respectively.

Model	Dev Set		Test Set	
	Acc.	**F1**	**Acc.**	**F1**
LR	68.5	81.4	68.6	81.4
SVM	71.2	83.4	70.7	83.3
LSTM	69.0 ±0.9	83.9 ±0.1	68.9 ±0.3	83.2 ±0.2
LSTM$_{reg}$	71.2 ±0.5	83.9 ±0.3	70.8 ±0.7	83.6 ±0.5
KimCNN	69.9 ±0.2	83.3 ±0.3	68.8 ±0.1	82.7 ±0.1
XML-CNN	72.9 ±0.4	84.1 ±0.2	71.7 ±0.7	83.5 ±0.3
BERT$_{base}$	74.3 ±0.6	85.5 ±0.4	73.6 ±1.0	85.1 ±0.5
BERT$_{large}$	75.1 ±3.9	85.9 ±1.9	74.4 ±2.7	85.3 ±1.4
Longformer	74.4 ±0.8	85.6 ±0.5	73.9 ±0.8	85.5 ±0.5
BioBERT	75.0 ±0.5	86.3 ±0.2	75.2 ±0.7	86.2 ±0.6

Table 2: Performance for each model expressed as *mean ± standard deviation* across three training runs.

We evaluate the data efficiency and generalizability of these models as essential features of any system prepared to deal with an urgent situation like the current health crisis.

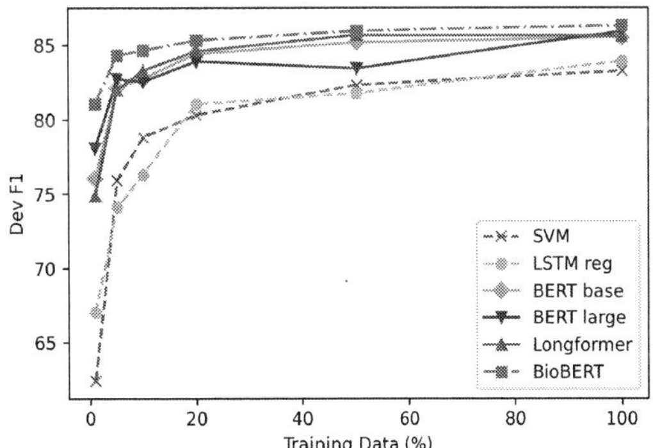

Figure 1: Data efficiency analysis.

All pre-trained language models tested are impressively data efficient, with BioBERT achieving an F1 score only 4 points below its maximum score using only 1% of the training data.

	Acc.	**F1**
SVM	29.0	62.8
LSTM$_{reg}$	32.7 ±1.5	67.7 ±0.7
Longformer	41.3 ±6.4	70.0 ±2.9
BioBERT	36.0 ±7.8	69.7 ±2.8

Table 3: Performance on the CORD-19 Test Set expressed as *mean ± standard deviation* across three training runs.

From Table 3, we can see that performance drops significantly on the CORD-19 test set which does not mention COVID-19. This shows that more work needs to be done for these models to be immediately useful in future health emergencies.

Finally, we explore 50 errors made by the best performing models on LitCovid documents and find that they often (1) correlate certain labels too closely together and (2) fail to focus on discriminative sections of the articles; both of which are important issues to address in future work. Both data and code are available on GitHub [1].

---

[1] `https://github.com/dki-lab/`
`covid19-classification`

Acknowledgments

This research was sponsored in part by the Ohio Supercomputer Center (Center, 1987). The authors would also like to thank Lang Li and Tanya Berger-Wolf for helpful discussions.

References

Ashutosh Adhikari, Achyudh Ram, Raphael Tang, and Jimmy Lin. 2019a. Docbert: Bert for document classification. *ArXiv*, abs/1904.08398.

Ashutosh Adhikari, Achyudh Ram, Raphael Tang, and Jimmy Lin. 2019b. Rethinking complex neural network architectures for document classification. In *NAACL-HLT*.

Simon Baker. 2017. Corpus and Software.

Simon Baker, Ilona Silins, Yufan Guo, Imran Ali, Johan Högberg, Ulla Stenius, and Anna Korhonen. 2016. Automatic semantic classification of scientific literature according to the hallmarks of cancer. *Bioinformatics*, 32 3:432–40.

Iz Beltagy, Matthew E. Peters, and Arman Cohan. 2020. Longformer: The long-document transformer. *arXiv:2004.05150*.

Ohio Supercomputer Center. 1987. Ohio supercomputer center.

Q. Chen, A. Allot, and Z. Lu. 2020. Keep up with the latest coronavirus research. *Nature*, 579(7798):193.

Jacob Devlin, Ming-Wei Chang, Kenton Lee, and Kristina Toutanova. 2019. Bert: Pre-training of deep bidirectional transformers for language understanding. *ArXiv*, abs/1810.04805.

Jingcheng Du, Qingyu Chen, Yifan Peng, Yang Xiang, Cui Tao, and Zhiyong Lu. 2019. Ml-net: multi-label classification of biomedical texts with deep neural networks. *Journal of the American Medical Informatics Association : JAMIA*.

Ruihua Fang, Gary Schindelman, Kimberly Van Auken, Jolene Fernandes, Wen J. Chen, Xiaodong Wang, Paul Davis, Mary Ann Tuli, Steven J. Marygold, Gillian H. Millburn, Beverley Matthews, Haiyan Zhang, Nick Brown, William M. Gelbart, and Paul W. Sternberg. 2011. Automatic categorization of diverse experimental information in the bioscience literature. *BMC Bioinformatics*, 13:16 – 16.

Xiangying Jiang, Martin Ringwald, Judith A. Blake, Cecilia N. Arighi, Gongbo Zhang, and Hagit Shatkay. 2019. An effective biomedical document classification scheme in support of biocuration: addressing class imbalance. *Database: The Journal of Biological Databases and Curation*, 2019.

Yoon Kim. 2014. Convolutional neural networks for sentence classification. In *EMNLP*.

Jinhyuk Lee, Wonjin Yoon, Sungdong Kim, Donghyeon Kim, Sunkyu Kim, Chan Ho So, and Jaewoo Kang. 2019. BioBERT: a pre-trained biomedical language representation model for biomedical text mining. *Bioinformatics*.

Jingzhou Liu, Wei-Cheng Chang, Yuexin Wu, and Yiming Yang. 2017. Deep learning for extreme multi-label text classification. *Proceedings of the 40th International ACM SIGIR Conference on Research and Development in Information Retrieval*.

Fabian Pedregosa, Gaël Varoquaux, Alexandre Gramfort, Vincent Michel, Bertrand Thirion, Olivier Grisel, Mathieu Blondel, Gilles Louppe, Peter Prettenhofer, Ron Weiss, Vincent Dubourg, Jacob VanderPlas, Alexandre Passos, David Cournapeau, Matthieu Brucher, Matthieu Perrot, and Edouard Duchesnay. 2011. Scikit-learn: Machine learning in python. *J. Mach. Learn. Res.*, 12:2825–2830.

Anthony Rios and Ramakanth Kavuluru. 2015. Convolutional neural networks for biomedical text classification: application in indexing biomedical articles. *ACM-BCB ... ... : the ... ACM Conference on Bioinformatics, Computational Biology and Biomedicine. ACM Conference on Bioinformatics, Computational Biology and Biomedicine*, 2015:258–267.

Lucy Lu Wang, Kyle Lo, Yoganand Chandrasekhar, Russell Reas, Jiangjiang Yang, Darrin Eide, Kathryn Funk, Rodney Michael Kinney, Ziyang Liu, William. Merrill, Paul Mooney, Dewey A. Murdick, Devvret Rishi, Jerry Sheehan, Zhihong Shen, Brandon Stilson, Alex D. Wade, Kuansan Wang, Christopher Wilhelm, Boya Xie, Douglas M. Raymond, Daniel S. Weld, Oren Etzioni, and Sebastian Kohlmeier. 2020. Cord-19: The covid-19 open research dataset. *ArXiv*, abs/2004.10706.

Pengcheng Yang, Xu Sun, Wei Li, Shuming Ma, Wei Wu, and Houfeng Wang. 2018. Sgm: Sequence generation model for multi-label classification. In *COLING*.

Enabling Low-Resource Transfer Learning across COVID-19 Corpora by Combining Event-Extraction and Co-Training

Alexander Spangher, Nanyun Peng, Jonathan May, Emilio Ferrara
Information Sciences Institute, University of Southern California, Marina del Rey, USA
{spangher, npeng, jonmay, ferrarae}@isi.edu

Abstract

Social-science investigations can benefit from a direct comparison of heterogenous corpora: in this work, we compare U.S. state-level COVID-19 policy announcements with policy discussions on Twitter. To perform this task, we require classifiers with high transfer accuracy to both (1) classify policy announcements and (2) classify tweets. We find that co-training using event-extraction views significantly improves the transfer accuracy of our RoBERTa classifier by 3% above a RoBERTa baseline and 11% above other baselines. The same improvements are not observed for baseline views. With a set of 576 COVID-19 policy announcements, hand-labeled into 1 of 6 categories, our classifier observes a maximum transfer accuracy of .77 f1-score on a hand-validated set of tweets. This work represents the first known application of these techniques to an NLP transfer learning task and facilitates cross-corpora comparisons necessary for studies of social science phenomena.

State-Level Policy	Label
The Governor issued a "Stay Safe, Stay Home" Directive.	Public Space Restriction
The Governor announced [loans] to provide relief for restaurants, bars, businesses.	Economic Measures
The Indiana State Department of Health announced new testing for COVID-19.	Healthcare

Tweets	Label
Can You Cancel Your Flight Because of the Coronavirus?	Travel Restrictions
Everyone should start wearing face masks in crowded public places.	Healthcare
What will it take for restaurants and ALL non-essential businesses to close?	Public Space Restriction

Table 1: Sample Policy Announcements (top) and Tweets (bottom) that we seek to classify, with labels.

1 Introduction

During the initial stages of the COVID-19 crisis, the U.S. lacked a centralized political response and a cultural familiarity with pandemics. From a social science perspective, research into COVID-19 policies and conversations around policy has consequences for both citizens and policy-makers as: (1) the extent of citizens' adherence to policy is often determined by awareness, and (2) in a democracy, policy-makers aim to produce policy that citizens are likely to support.

To lay the methodological groundwork for a such a comparative study of policy and conversation, we examine methods for cross-corpora transfer learning across COVID-19 government policy statements and COVID-19 tweets. Such a task is challenging because the style and intent of each corpus differs while the events discussed do not; i.e.,

the vocabulary is divergent in some respects but similar in others. Previous transfer learning work involving Twitter has stayed within tweet corpora: i.e. multi-lingual tasks (Levy and Yang Wang) or tweets-to-comment tasks (Tian et al., 2020) where the style of the transfer task is similar. Comparatively less work has focused on comparing a non-social media corpora with Twitter.

To overcome the stylistic differences between language in government policy documents and tweets, we utilize event-extraction and co-training, achieving a 11% improvement over baseline. We hypothesize that event-extraction allows us to focus on similarities between corpora while co-training allows us to impart signal from differences.

2 Dataset

We use a dataset of roughly 137 million tweets that researchers collected by analysing variations of the COVID-19 hashtag (Chen et al., 2020a) and 2,100 state-level government policy announcements that we scraped and parsed from the National Governors Association website.[1]

Each tweet datum consists of the full text of the tweet, the author's name, and the date it was sent, among other features. Each governor's announcement datum consists of the state of the governor, the text of the announcement, and the date it was announced.

We describe first our method for labeling state-level policy announcements. Then we describe our method for filtering COVID tweets to policy-related tweets. Finally, we discuss the models that we tested in order to classify state policy and tweets and the steps we took to increase transfer accuracy.

2.1 Labeling Policy Announcements

We hand-label the first 576 governor announcements that we collect, each into one of 6 categories. We choose categories to help us identify different treatments for downstream variables (for analysis in upcoming work, noted in Section 1). Additionally, the top-level labels independently correspond to treatment categories outlined on the National Governor Associations website[1].

Two annotators together duplicately label 120 of the announcements without conferring throughout the process. On these 120 labels, we report an inter-annotator agreement of $\kappa = .76$. The classes are imbalanced and the number of labels in each class is shown in Table 2.

The hierarchical categories encompass broad different policy-types. For instance, the "Government Preparedness" category corresponds to measures policymakers took to prepare their governments for the crisis, including: "Summoning the National Guard", and "establishing a Task Force". Numerous policies do not neatly fit into a subcategory: thus, while a rule-based approach was considered to identify top-level categories, we decided a classification-based approach was necessary.

2.2 Filtering Policy-Related Tweets

We filter tweets to those that use similar language as the policy announcements in order to identify policy-relevant tweets. To do this, we derive a

Policy Announcement Type	Count
Public Space Restriction	185
Government Preparedness	161
Economic Measures	80
Healthcare	75
Remote Working Policies	22
Travel Restriction	18
Total	576

Table 2: Number of hand-labeled policy announcements in each class.

Processing	Example
Text	Governor Ivey issued a state of emergency for Alabama .
Events	Ivey issued state of emergency .
Text & Events	Governor Ivey issued a state of emergency for Alabama. Ivey issued state of emergency .

Table 3: A sample policy announcement processed using the 6 data-processing approaches used in our classification experiments. Full-text was processed using event-extraction and lemmatization.

bag-of-words representation for each policy announcement and for each tweet. For each tweet, we calculate its pairwise cosine similarity with all the policy announcements. If an announcement exists with a similarity above .35 to the tweet, then we consider it a policy-related tweet. We choose this threshold after cross-validation: our annotators hand-validated a set of 120 tweets as policy-related or not.[2] Our method achieves an f1-score of .78 for identifying policy-related tweets[3].

2.3 Preprocessing

We preprocess our data before classifying in six different ways, shown in Table 3. As a baseline, we consider not preprocessing – simply classifying the raw text of the announcement or tweet. Another method we consider is the lemmatization of the sentence.[4] Lemmatization was considered after noting that policy announcements and tweets have significantly different distributions over the tense

---

[2] We additionally set our vocabulary inclusion criteria after cross-validation as well. Our inclusion criterion for words was that each had to appear in at least 2 documents and not more 40% of documents in the policy corpus. This resulted in $1,021$ words.

[3] Note: this evaluation is separate from the classification task described in later sections.

[4] Lemmas are derived using https://spacy.io/.

	Present/Future	Past
Tweets	0.92	0.08
Policy	0.63	0.37

Table 4: Percentage of sentences in tweets and policy that are in past or present/future tense.

of their sentences, as shown in Table 4[5]. While lemmatization should mitigate that difference, it does not measurably improve the accuracy of our classifiers. Additionally, for the LogisticRegression classifier, we experimented with different vocabulary thresholds before choosing an optimal set, based on performance.

Another set of features we consider are extracted events: we extract event arguments – agents and patients – and anchors using a BERT + BiLSTM neural architecture (Han et al., 2019), as well as the lemmatized version of these extracted events. We consider event-extraction after observing that tweets are significantly more likely to contain opinionated text – policy text has a median subjectivity of .23 while tweet text has a median subjectivity of .33.[6] We hypothesize event-extraction can help transfer accuracy by abstracting the content of tweets from opinions.

3 Methodology

3.1 Classification

We test two classifiers: Logistic Regression on a TF-IDF normalized[7] bag-of-words representation of each input document, and a pretrained RoBERTa-base model. We use Logistic Regression because it is a fast and interpretable baseline. We consider optimization across a range of l_2-norm constraints on the coefficient sizes and we optimize on held-out training data. We report the top-performing model in our results.

We use RoBERTa because it is a language model that has been shown to generalize well across transfer tasks (Raffel et al., 2019). Importantly, a pretrained language model is ideal when the transfer corpus might contain vocabulary words not contained in the original corpus. We use the

View	Example
Full Text	Shanghai Disneyland closed during Lunar New Year due to coronavirus!!
#1: Events	closed Shanghai Disneyland
#2: Text without Events	during Lunar New Year due to coronavirus!!

Table 5: **Co-training preprocessing:** A sample tweet with views used for co-training. **View #1** is the extracted events, **View #2** is the text with event words removed. The labels generated from one view are iteratively added to another view's training set. Then, the full text along with all added labels are used to train the final classifier. Not shown but tested as baseline views are: "noun-phrases", "verb-phrases", "random words".

RoBERTa-base pretrained model[8] for our classification task, although we acknowledge that a more corpus-specific pretraining, like a Twitter-specific pretraining or a law-specific pretraining might achieve higher accuracy (Nguyen et al., 2020).

3.2 Co-Training

Additionally, we test co-training as a method for increasing transfer accuracy. As formulated by Blum and Mitchell (1998), co-training is a method for increasing the accuracy of a classifier by using labels generated by other classifiers with different "views" of the data (i.e. non-overlapping feature sets). Previous work has found co-training advantageous in transfer learning tasks (Wan, 2009).

The views we use are shown in Table 5. For View #1, we extract events using the method as described in Section 3.1, and for View #2, we leave the text without events as the other view. To test the efficacy of events, we consider additional baseline views: noun-phrases, verb-phrases, and random words. For each of the baseline views, we extract the linguistic component as one view and leave the rest of the sentence as the other.

We cycle iteratively between views, where for each view, we train a classifier, use it to label the top k most confident unlabeled datapoints for each class, and add them to the training set (we test different values of k using heldout data, and choose $k = 15$, or 30 datapoints from both views. This is roughly 5% the size of the original dataset[9]). At

---

[5]We determined sentence tense by checking whether the root of the sentence was a past-tense verb, or if any of the roots children were past-tense with an auxiliary dependency.

[6]Derived using the Python package TextBlob.

[7]TF-IDF refers "term-frequency inverse document frequency" Words are counted by their frequency in the document, normalized by their overall appearance in the corpus.

[8]Provided by huggingface.co.

[9]In the original co-training paper, authors added 2% the size of the original dataset (Blum and Mitchell, 1998)

each iteration, we add the newly labeled data under one view to the training set used by the other view (each view's classifier only ever sees data labeled by the other view). Also, we test classifier accuracy trained on the full text with all additional labeled points.

4 Results

4.1 Experimental Setup

We measure classifier accuracy on two principal tasks: (1) the baseline classification task on government policy data and (2) the transfer task on Twitter data.

To report the baseline task, we show 5-fold cross validation accuracy on the labeled dataset, which consists of 576 hand-labeled policy announcements. To report on the transfer task, we randomly select and label a batch of 310 policy-related tweets. We use these tweets to validate our models' output and show confidence using bootstrapped samples.

4.2 Accuracy of Classifiers: No Co-training

In Figure 1a we show micro f1-scores across the 5-fold validation tests of our classifiers across the 6 labeled classes. For the **RoBERTa** classifier, event-extraction has only a marginal effect on held-out accuracy: only for the **RoBERTa Text** classifier does appending events increase the median micro-f1 score. The highest-performing classifier is **RoBERTa Text & Events**, but the addition of events does not significantly improve the classifier above **RoBERTa Text**.

Next, we test how well our classifiers transfer to labeling the Twitter data. Figure 1b shows the micro-f1 scores on this validation data over 500-bootstrapped samples. Similar to the previous analysis, adding event information explicitly to the classification task does not significantly change the accuracy. Taken together, these two analyses suggest that event information is not useful for increasing classification accuracy without co-training.

4.3 Co-Training

To test the impact of co-training on our transfer learning task, we used co-training in two settings: (1) we add co-trained labels to new data from the policy dataset (which is in the same domain as the training set), (2) we add co-trained labels to new data from the Twitter dataset (which is from a different domain than the training set).

Figure 2 shows the results of these two experiments using Logistic Regression as a classifier and Figure 3 shows the results using RoBERTa as a classifier. In each figure, we report the accuracy of our classifiers at each iteration of data augmentation on each view as well as on the full-training text.

As shown in Figure 2a, when co-training is tested on heldout data from the training set, it has a positive effect for the Logistic Regression classifier, increasing the median f1-score across 5-folds from .82 to .84. The result is not significant across the IQR interval (25th percentile to 75th percentile of held-out runs). On the other hand, as shown in Figure 3a, co-training had no effect on the RoBERTa classifier, perhaps indicating that RoBERTa already reached a ceiling on the amount of signal it derived from the original training set.

We had hypothesized that co-training could be successful in helping our models generalize to different corpora. As shown in Figure 2b, co-training had a negligible effect when applied to the transfer corpus using Logistic Regression. However, as shown in Figure 3b, co-training had a positive effect, increasing the validation accuracy from .745 to .774, an increase that was significant across a 500-bootstrapped sample.

We hypothesized that event-information was particularly useful for co-training as it added the most meaning to one view (View #1) while also being conditionally independent of the other (View #2). To test this hypothesis, we compared different views, shown in Figure 4. Figure 4a shows the different views tested on the baseline classification task. While co-training with events is on par with co-training with noun-phrases, events performs better across folds, indicating better generalization. Figure 4b shows the different views tested on the transfer task. Here, co-training with events is the clear winner, and the only view to increase the accuracy across tasks.

To explore why co-training might increase the accuracy of our classifiers, we examined the size of the vocabulary in training documents across co-training iterations in Figure 5. As shown in Figure 5a, the size of the vocabulary was relatively constant when policy documents were added to the labeled set during co-training whereas when tweets were added to the labeled set, the vocabulary continued increasing linearly with the number of iterations.

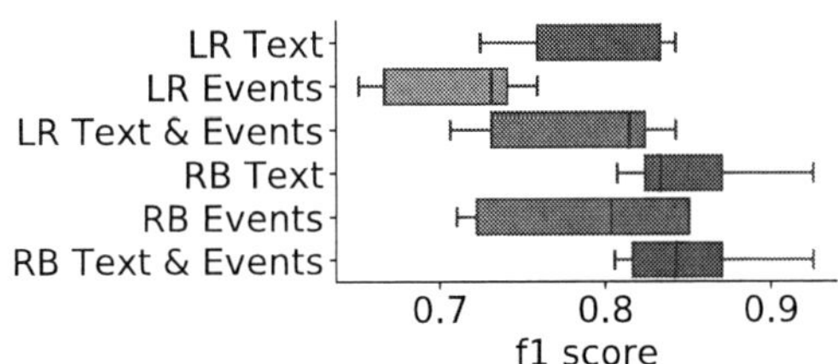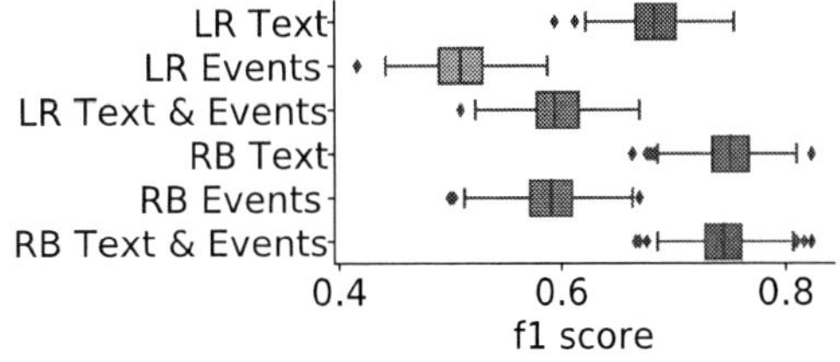

(a) **Classifier Accuracy on 5-fold holdout annotated policy data.** The highest score across 5 folds is for **Text & Events** with a RoBERTa classifier, achieving a median micro f1 score of .85.

(b) **Classifier Accuracy on 310 hand-validated tweets (500 bootstrapped sample).** The highest score is for **Text** with a RoBERTa classifier, acheiving a median micro-f1 score across bootstrapped samples of .75.

Figure 1: **Classifier Accuracy** on (a) training corpus and (b) transfer corpus. The first three rows show different text preprocessing using Logistic Regression classifier (LR); the last three show the RoBERTa classifier (RB).

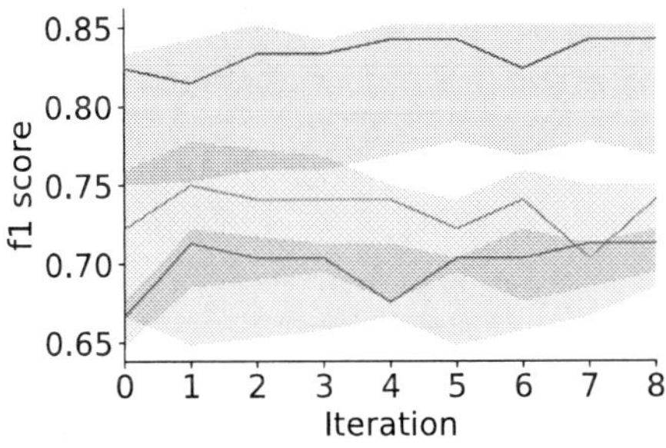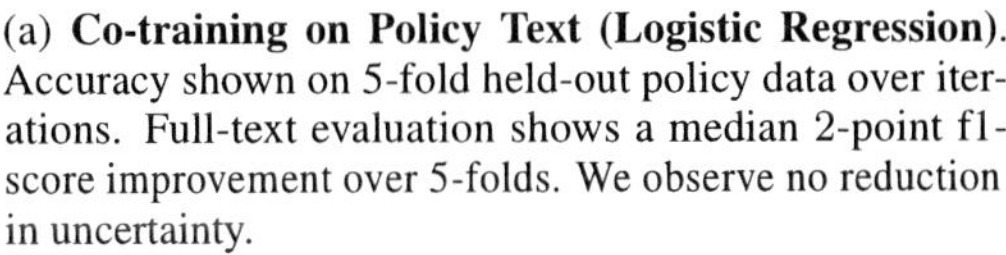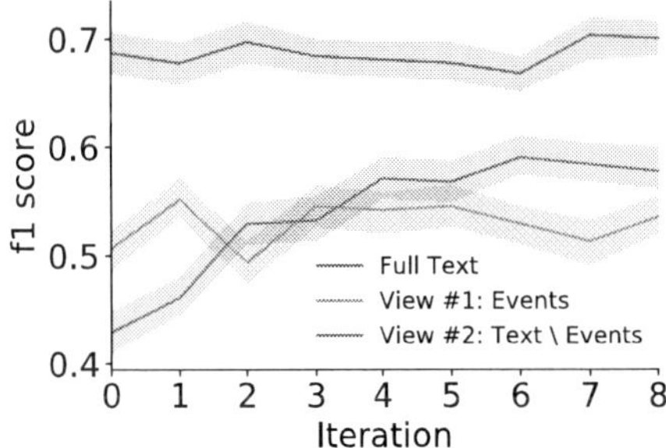

(a) **Co-training on Policy Text (Logistic Regression).** Accuracy shown on 5-fold held-out policy data over iterations. Full-text evaluation shows a median 2-point f1-score improvement over 5-folds. We observe no reduction in uncertainty.

(b) **Co-training on Tweets (Logistic Regression).** Validation accuracy shown on 310 hand-labeled tweets (500 bootstrapped sample). Full-text evaluation shows a median 1-point f1-score improvement at iteration 8 from baseline at iteration 0, from .69 to .70.

Figure 2: **Co-training accuracy** across 8 iterations of co-training data augmentation for (a) training corpus and (b) transfer corpus. Co-training was performed using a **Logistic Regression** classifier on two views of the data: results shown for *Full-Text* (blue), *View #1* (Orange) and *View #2* (Green). The training accuracy for iteration 0 represents the baseline dataset (i.e. *Full Text* at iteration 0 corresponds to *LR Text* in Figure 1). Iteration i represents $i \times k \times l$ dataset augmentation for training, where $k = 15$ is the number of co-training datapoints added per turn per class, and l is the number of classes. *Full-Text* is run at each turn on the entire dataset compiled by the co-training views, as an evaluation. Only the two views contribute labeled data.

This had an effect on the words considered high-signal by our Logistic Regression classifier (i.e. with the highest absolute-valued coefficients). In Table 6, we show the top words at the beginning of co-training, when the training set included only policy documents. In Table 7, we show high-signal words at the end of co-training (iteration 8), when our training corpus contained 720 tweets in addition 576 policy documents.

We also examine the percentage of the total high-signal words that were added – defined as those with the largest k coefficients in the Logistic Regression model. As shown in Figure 5b, the percentage of of high-signal words not in the policy document corpus increases nonlinearly with the number of co-training iterations. The maximum amount of high-signal words across classes not in the policy documents occurs at the sixth iteration, when 31.5% of the top 100 high-signal words across classes were not in the policy documents. This corresponds to a decrease in accuracy on the sixth iteration in Figure 2b.

5   Discussion

Co-training increases the signal of a downstream classifier when the dataset is independent, the views are conditionally independent given the label, and when the classifiers do not agree on the same datapoints (Krogel and Scheffer, 2004).

We can assume that the datapoints are indepen-

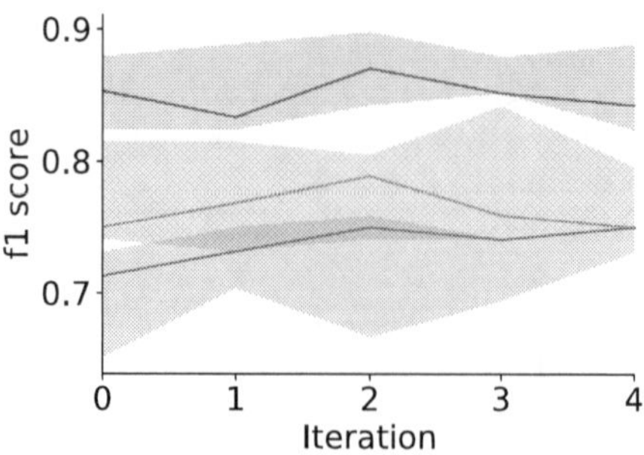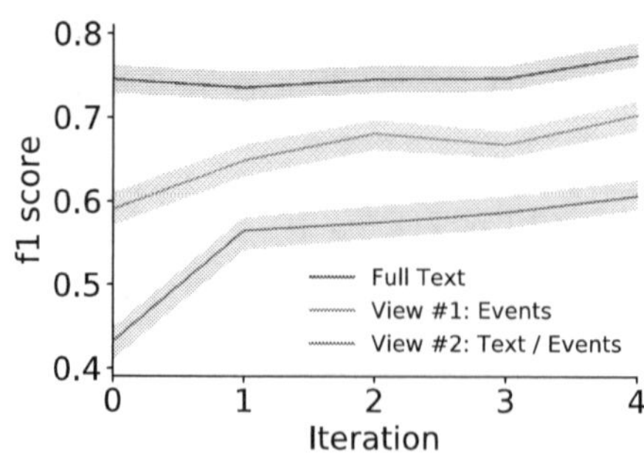

(a) **Co-training on Policy Text (RoBERTa)**. Accuracy shown on 5-fold held-out policy data over 4 iterations. We observe a maximum median f1 score of .87 occurs at iteration 2 although statistically insignificant given IQR.

(b) **Co-training on Tweets (RoBERTa)**. Validation accuracy shown on 310 hand-labeled tweets (500 bootstrapped sample). Full-text evaluation shows a 3-point improvement at iteration 4 from baseline (iteration 0), .74 to .77.

Figure 3: **Co-training accuracy** across 4 iterations of co-training data augmentation for (a) training corpus and (b) transfer corpus. Co-training was performed using RoBERTa classifier.

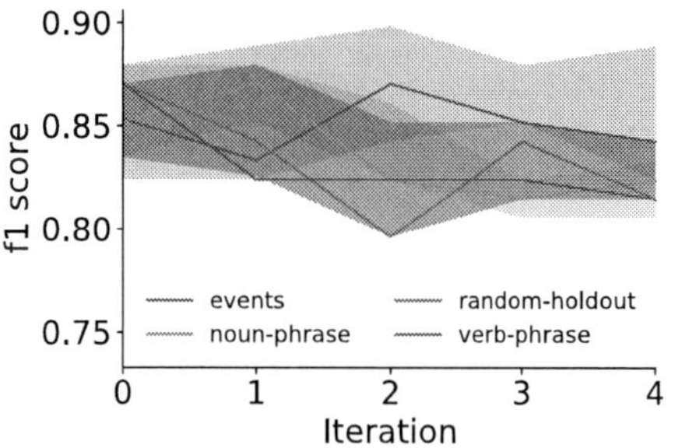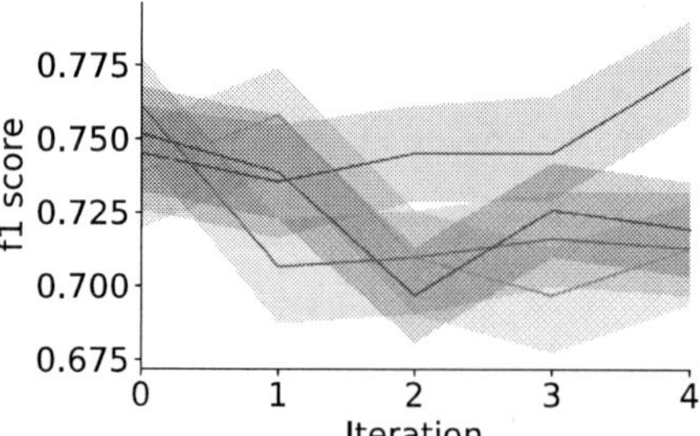

(a) **Co-training on Policy Text (RoBERTa), Alternate Views**. Accuracy shown on 5-fold held-out policy data over 4 iterations. Maximum IQR f1 score (median=.87) occurs with events at iteration 2.

(b) **Co-training on Tweets (RoBERTa), Alternate Views**. Validation accuracy shown on 310 hand-labeled tweets (500 bootstrapped sample). Events View significantly outperforms other views at iteration 4.

Figure 4: **Co-training accuracy, different views** across 4 iterations of co-training data augmentation for (a) training corpus and (b) transfer corpus. Co-training was performed using RoBERTa classifier.

dent between corpora, as we do not observe direct tweeting about specific state-level policy. Additionally, we observe a low degree of agreement on the same datapoints – across the iterations, we observe the two classifiers recommending the same new sample with the same tag less than 5% of the time, across the two models. In these cases, we randomly choose one of the views to assign the sample to, and drop it from the other. Finally, although we are strictly partitioning the words into one view or the other, we are not as confident about conditional independence between the views, as there may still be structural and syntactic dependence between the two views that we did not account for.

By tracking the accuracy of each view as well as the full-text accuracy, we can see that in all cases (i.e. both co-training setups under both models) both views gain the signal in the first round of co-training. In all but one case, **View #2: Text \ Events** has both the lowest accuracy and the greatest improvements in accuracy in the early iterations – the exception being the transfer task and Logistic Regression (Figure 2b), where **View #1: Events** gains more accuracy in the first iteration.

Interestingly, in all cases, the full-text output decreases in accuracy after the first turn. At all rounds, the vocabulary of our interpretable model, Logistic Regression, increases, according to Figure 5a. An increasing vocabulary means that co-training was adding documents to our training set that used words that were different from our original training set with enough frequency to be included. According to Figure 5b, in the first round of co-training, the high-signal non-policy vocabulary increases at nearly the highest rate (the highest being at round 6). We hypothesize, based on this vocabulary anal-

Econ. Meas.	Gov. Prep.	Healthcare	Public Space Rest.	Travel Rest.	Remote Work
childcare	national	division	schools	employees	employees
meals	guard	testing	march	asking	meetings
unemployment	emergency	health	gatherings	travel	telework

Table 6: Top words by coefficient in Logistic Regression model at co-training iteration 0 (i.e. Logistic Regression trained on 576 policy documents).

Econ. Meas.	Gov. Prep.	Healthcare	Public Space Rest.	Travel Rest.	Remote Work
sick	guard	healthcare	schools	university	person
workers	national	health	people	international	employees
unemployment	emergency	testing	gatherings	travel	meetings

Table 7: Top words by coefficient in Logistic Regression model at co-training iteration 8 (i.e. Logistic Regression trained on 576 policy documents and 720 co-trained tweets). Shaded cells are words not considered high-signal in each class before co-training.

ysis, that the first turn is when the domain expands the most without enough labels to overcome this increase in input dimensionality.

Our vocabulary analysis suggests an additional rationale for why co-training delivered a far greater effect with RoBERTa – three times the improvement – than it did with Logistic Regression. While co-training using Logistic Regression may have increased the label set available, it also increased the new words that the model had to learn. It could be that the amount of signal available in the labels could not overcome the sparsity introduced by the vocabulary expansion. In contrast, because we utilized pretrained RoBERTa, the new vocabulary added by the datapoints did not add a corresponding level of sparsity. This suggests that a more domain-specific pretrained model, like a Twitter-specific RoBERTa (Nguyen et al., 2020), might have even greater benefits from co-training, but we leave that to future work.

We also leave to future work an exploration of further views that could increase co-training accuracy. Further engineering tricks, such as selective lemmatizing, might perform well as views. However, as indicated in Figure 4, event-extraction is a particularly useful view for imparting signal, relative to the baseline views we considered – baselines which have been, in fact, used in the literature (Pierce and Cardie, 2001). While it's not immediately clear why this is the case, we hypothesize that extracting events as one view gives us the clearest conditional independence of views, which is necessary for co-training to be effective. Interestingly, both the "noun-phrase" baseline and the "verb-phrase" baseline degrade in performance over

time for the transfer task (Figure 4b). It may be that these two tasks separate the views similarly: i.e., what is left over when extracting noun-phrases is mostly verb-phrases, and vice-versa. As shown in (Nigam and Ghani, 2000), if the accuracy of even the most confident co-training labels degrades over time, then the performance of the co-training labeled set will also decline. It might be, then, that none of the views contained enough signal to make confident predictions.

Overall, in contrast to Figure 1 where **Text & Events** had a mixed effect on the classification accuracy, we show in this paper that co-training using event-extraction to parse sentences into two views can be useful in adding signal to a transfer-learning task.

6   Related Works

6.1   Policy and Twitter Analysis for Coronavirus

Emerging work has sought to explore the effects of various policy approaches to the coronavirus outbreak both nationally (Stock, 2020) and in specific locales (Friedson et al., 2020). Such work uses a limited number of curated policies to build up treatment sets: the dataset of treatments is composed of indicator variables and requires manual annotation and collation to collect. Because we seek to generalize and expand the treatment dataset (state-level policies), and explore the conversations around such treatments, we need to associate treatment labels with specific spans of text (governor policy announcements) to train classifiers.

Recent work has also sought to study the con-

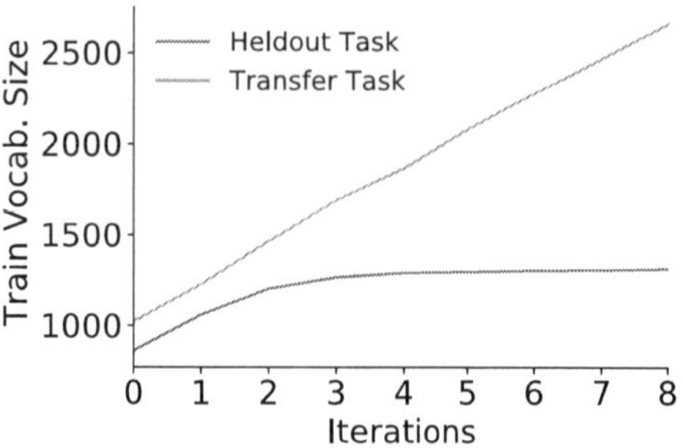

(a) **Size of Vocabulary** given max/min constraints for Logistic Regression over co-training iterations.

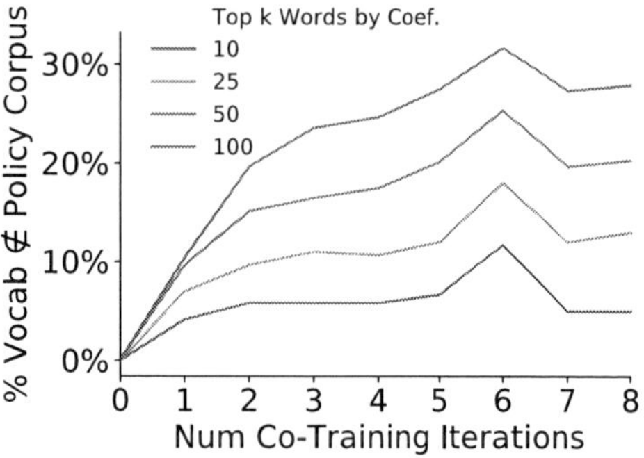

(b) **Percentage of Vocabulary** carrying high signal that is not in the policy corpus, i.e. exists only in the Twitter corpus. We define "high signal" as the top k largest coefficients in the Logistic Regression model.

Figure 5: Increasing size of vocabulary for Transfer task shows increasingly diverse training data is being included in the task at each iteration relative to the Heldout Task, where the vocabulary stays constant over time.

versations around coronavirus on Twitter. Early work centered on collecting the dataset (Chen et al., 2020a) and followup focused on characterizing the conversations occurring on Twitter (Chen et al., 2020b). Such work involves unsupervised content analysis largely based on hashtag counts and topic modeling. Unsupervised analysis is necessary in cases where labeled data is unavailable. Additional followup work has focused on characterizing conspiracy-related conversations encouraged by bots on Twitter (Ferrara, 2020). This work uses pretrained bot-detection models to identify bots, and unsupervised methods to summarize the content. Our work differs from this work in that we seek to analyse a specific part of the Twitter conversation – the policy-related conversation. As such, we need some notion of what specific threads of policy exist.

6.2 Cross-Domain Text Classification

This work is in line with cross-domain text classification, which has been studied widely. As such, the summary given here is limited, however, much work in this field focuses on areas where labeled data is plentiful, such as sentiment classification across corpora. For example, in early work (Blitzer et al., 2007), authors used regularized linear classifiers to enforce finding common word-features across corpora. They collected positive and negative Amazon product reviews in four categories and gathered $O(10, 000)$ labeled datapoints. Recent work using the same dataset utilizes shared-private encoders has leveraged advances in neural networks to train high-accuracy models (Wu et al., 2019).

Both of these works seek to use novel classification architectures that are fine-tuned to the task of multi-domain learning: the first uses classical techniques (i.e. regularization) to enforce learning similarities between domains while the other uses neural architectures to achieve the same goal. However, both enjoy the advantage of relatively plentiful data. And, because both are tested only on held-out data, neither needs to generalize beyond the labeled dataset. Our work, on the other hand, requires a generalizability to unseen future policy text and tweets, and we have a comparatively smaller dataset.

Previous work has addressed such problems using co-training, which is the meta-approach that we take. In the original co-training paper, authors used only 12 labeled data to correctly categorize 95% of 788 web pages (Blum and Mitchell, 1998). Researchers have applied co-training to pure-text classification as well, using noun-phrases to split views (Pierce and Cardie, 2001) as well as heuristics – for instance, (Denis et al., 2003) classifies scientific articles using "author information" as one view and body text as another. The limitations of co-training have been explored: in addition to limitations mentioned in previously, researchers show that when classes are imbalanced, performance degrades (Kiritchenko and Matwin, 2011). To our knowledge, our work is the first to use event-extraction in co-training for a downstream application.

7 Conclusion

Policy makers can benefit from understanding public conversation around policy: public conversation can help them (1) understand what policies would be supported and (2) track awareness of policies intended to promote social wellbeing. Such work is particularly important in the COVID-19 crisis, which both dominated the national conversation and involved policy responses that U.S. citizens were heretofore unaccustomed to.

In this work, we lay the methodological groundwork for such studies of COVID policy. We classified both policy responses and tweets, and have shown a significant improvement from our Logistic Regression classifier (which achieved .69 f1-score on our transfer task), to our co-training RoBERTa classifier (which achieved a .77 f1-score, with only 576 original labeled examples). Such an improvement allows us to compared policies and tweets, effectively normalizing for linguistic differences between the corpora. Such work paves the way for ongoing work in examining the interplay between policy and conversation. We hope that future work can leverage both our labeled data as well as our cross-domain class predictions to inform policy makers in COVID-prevention work.

References

John Blitzer, Mark Dredze, and Fernando Pereira. 2007. Biographies, bollywood, boom-boxes and blenders: Domain adaptation for sentiment classification. In *Proceedings of the 45th annual meeting of the association of computational linguistics*, pages 440–447.

Avrim Blum and Tom Mitchell. 1998. Combining labeled and unlabeled data with co-training. In *Proceedings of the eleventh annual conference on Computational learning theory*, pages 92–100.

Emily Chen, Kristina Lerman, and Emilio Ferrara. 2020a. Covid-19: The first public coronavirus twitter dataset. *arXiv preprint arXiv:2003.07372*.

Emily Chen, Kristina Lerman, and Emilio Ferrara. 2020b. Tracking social media discourse about the covid-19 pandemic: Development of a public coronavirus twitter data set. *JMIR Public Health and Surveillance*, 6(2):e19273.

Francois Denis, Anne Laurent, Rémi Gilleron, and Marc Tommasi. 2003. Text classification and co-training from positive and unlabeled examples. In *Proceedings of the ICML 2003 workshop: the continuum from labeled to unlabeled data*, pages 80–87.

Emilio Ferrara. 2020. What types of covid-19 conspiracies are populated by twitter bots? *First Monday*.

Andrew I Friedson, Drew McNichols, Joseph J Sabia, and Dhaval Dave. 2020. Did california's shelter-in-place order work? early coronavirus-related public health effects. Technical report, National Bureau of Economic Research.

Rujun Han, Qiang Ning, and Nanyun Peng. 2019. Joint event and temporal relation extraction with shared representations and structured prediction. *arXiv preprint arXiv:1909.05360*.

Svetlana Kiritchenko and Stan Matwin. 2011. Email classification with co-training. In *Proceedings of the 2011 Conference of the Center for Advanced Studies on Collaborative Research*, pages 301–312. IBM Corp.

Mark-A Krogel and Tobias Scheffer. 2004. Multi-relational learning, text mining, and semi-supervised learning for functional genomics. *Machine Learning*, 57(1-2):61–81.

Sharon Levy and William Yang Wang. Cross-lingual transfer learning for covid-19 outbreak alignment. *ACL 2020 Workshop NLP-COVID Submission*.

Dat Quoc Nguyen, Thanh Vu, and Anh Tuan Nguyen. 2020. Bertweet: A pre-trained language model for english tweets. *arXiv preprint arXiv:2005.10200*.

Kamal Nigam and Rayid Ghani. 2000. Analyzing the effectiveness and applicability of co-training. In *Proceedings of the ninth international conference on Information and knowledge management*, pages 86–93.

David Pierce and Claire Cardie. 2001. Limitations of co-training for natural language learning from large datasets. In *Proceedings of the 2001 Conference on Empirical Methods in Natural Language Processing*.

Colin Raffel, Noam Shazeer, Adam Roberts, Katherine Lee, Sharan Narang, Michael Matena, Yanqi Zhou, Wei Li, and Peter J Liu. 2019. Exploring the limits of transfer learning with a unified text-to-text transformer. *arXiv preprint arXiv:1910.10683*.

James H Stock. 2020. Data gaps and the policy response to the novel coronavirus. Technical report, National Bureau of Economic Research.

Lin Tian, Xiuzhen Zhang, Yan Wang, and Huan Liu. 2020. Early detection of rumours on twitter via stance transfer learning. In *European Conference on Information Retrieval*, pages 575–588. Springer.

Xiaojun Wan. 2009. Co-training for cross-lingual sentiment classification. In *Proceedings of the Joint Conference of the 47th Annual Meeting of the ACL and the 4th International Joint Conference on Natural Language Processing of the AFNLP: Volume 1-volume 1*, pages 235–243. Association for Computational Linguistics.

Haiming Wu, Yue Zhang, Xi Jin, Yun Xue, and Ziwen Wang. 2019. Shared-private lstm for multi-domain text classification. In *CCF International Conference on Natural Language Processing and Chinese Computing*, pages 116–128. Springer.

Self-supervised context-aware Covid-19 document exploration through atlas grounding

Dusan Grujicic,* Gorjan Radevski*, Tinne Tuytelaars, and Matthew B. Blaschko
ESAT-PSI, KU Leuven, Kasteelpark Arenberg 10, 3001 Leuven, Belgium
`{firstname.lastname}@esat.kuleuven.be`

Annotating and systematizing the increasing quantity of COVID-19 articles requires the expertise of physicians, and is cost-prohibitive. We focus on grounding such articles from the CORD-19 dataset (Wang et al., 2020) to locations in a 3D model of the human body, where the physical proximity of objects tends to reflect their semantic relatedness, allowing for a visual navigation. In a self-supervised manner, we use a mixture of organ term occurrences within a sentence as the indication of what the sentence denotes. A reference location for medical terms of interest is obtained from a 3D atlas of human anatomy based on the Visible Human male[1] and the Segmented Inner Organs[2] (Pommert et al., 2001; Höhne et al., 2001). To learn a context dependent grounding, we stochastically mask the organ related terms in a sentence during training and predict the coordinates within the human body. The loss represents the average of the Soft-min weighted distances between the prediction and the organ points, which emphasizes the most nearby target locations. We use BioBert (Lee et al., 2019) as our model, and obtain the results in Table 1, where *Regular* and *Masked* denote grounding of the regular sentences and sentences where all organ related words are substituted with a [MASK] token respectively, forcing the model to rely solely on the context. This indicates that the model is suitable for document exploration and retrieval through the 3D human atlas.

When performing text retrieval in the physical space, we can make a query by specifying a desired location in the human atlas, and directly observe the relationship between the embedded texts in a intuitive way. We build a tool (Fig. 1, Right) where

Method	Regular	Masked
Center	10.77 ± 0.10	10.77 ± 0.10
Frequency	9.49 ± 0.15	9.49 ± 0.15
BioBert (ours)	**0.21 ± 0.02**	**2.92 ± 0.08**

Table 1: Distance to the nearest voxel of the correct organ (in cm). Center - Predicts the center of the atlas, Frequency - Predicts the center of most frequent organ.

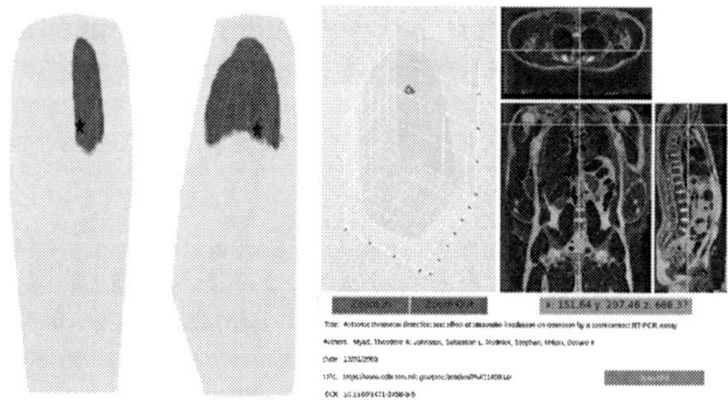

Figure 1: **Left**: Implicit lung reference grounding. **Right:** Point-cloud corpus visualization tool.

the 3D location is queried by a combination of 2D point selections on orthogonal cross-sections, while the nearest articles' 3D embeddings are shown on the left panel, where the user can zoom in and navigate between the closest suggestions. The tools and the code can be accessed at `https://github.com/gorjanradevski/macchina/`.

References

Höhne et al. 2001. A realistic model of human structure from the visible human data. *Methods of information in medicine*, 40(02):83–89.

Lee et al. 2019. Biobert: pre-trained biomedical language representation model for biomedical text mining. *arXiv:1901.08746*.

Pommert et al. 2001. Creating a high-resolution spatial/symbolic model of the inner organs based on the visible human. *MEDIA*, 5(3):221–228.

Wang et al. 2020. Cord-19: The Covid-19 open research dataset. *arXiv:2004.10706*.

*Equal contribution. Supported by KU Leuven Internal Funds (MACCHINA) and the Flemish Government under the Onderzoeksprogramma Artificiële Intelligentie (AI) Vlaanderen programme.

[1]`www.nlm.nih.gov/research/visible/`
[2]`www.voxel-man.com`

CODA-19: Using a Non-Expert Crowd to Annotate Research Aspects on 10,000+ Abstracts in the COVID-19 Open Research Dataset

Ting-Hao (Kenneth) Huang[1], Chieh-Yang Huang[1] Chien-Kuang Cornelia Ding[2],
Yen-Chia Hsu[3], C. Lee Giles[1]
[1]Pennsylvania State University, University Park, PA, USA
{txh710,chiehyang,clg20}@psu.edu
[2]University of California, San Francisco, CA, USA. Cornelia.Ding@ucsf.edu
[3]Carnegie Mellon University, Pittsburgh, PA, USA. yenchiah@andrew.cmu.edu

Abstract

This paper introduces **CODA-19**[1], a human-annotated dataset that codes the **Background, Purpose, Method, Finding/Contribution, and Other** sections of 10,966 English abstracts in the COVID-19 Open Research Dataset. CODA-19 was created by 248 crowd workers from Amazon Mechanical Turk within 10 days, and achieved labeling quality comparable to that of experts. Each abstract was annotated by nine different workers, and the final labels were acquired by majority vote. The inter-annotator agreement (Cohen's kappa) between the crowd and the biomedical expert (0.741) is comparable to inter-expert agreement (0.788). CODA-19's labels have an accuracy of 82.2% when compared to the biomedical expert's labels, while the accuracy between experts was 85.0%. Reliable human annotations help scientists access and integrate the rapidly accelerating coronavirus literature, and also serve as the battery of AI/NLP research, but obtaining expert annotations can be slow. We demonstrated that a non-expert crowd can be rapidly employed at scale to join the fight against COVID-19.

1 Introduction

As COVID-19 spreads across the globe, it's hard for scientists to keep up with the rapid acceleration in coronavirus research. Researchers have thus teamed up with the White House to release the COVID-19 Open Research Dataset (CORD-19) (Wang et al., 2020), containing 130,000+ related scholarly articles (as of August 13, 2020). The Open Research Dataset Challenge has also been launched on Kaggle to encourage researchers to use cutting-edge techniques to gain new insights from these papers (AI, 2020). To comprehend

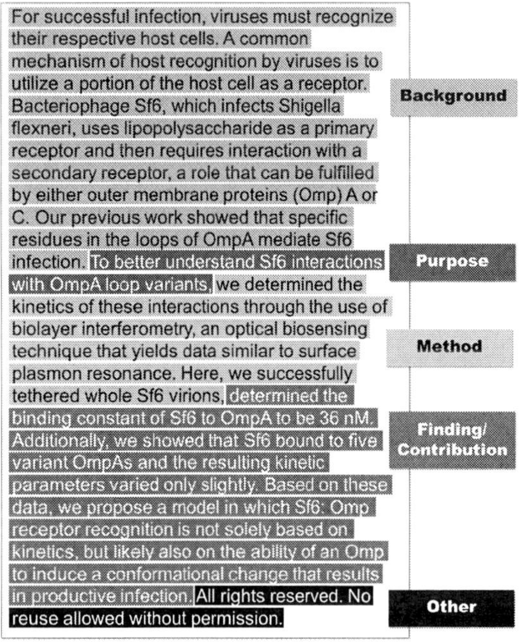

Figure 1: An example of the final crowd annotation for the abstract of (Hubbs et al., 2019).

the core arguments and contributions of a scientific paper effectively, it is essential to understand that most papers follow a specific *structure* (Alley, 1996), where aspects or components of research are presented in a particular order. A paper typically begins with the background information, such as the motivation to study and known facts relevant to the problem, followed by the methods that the authors used to study the problem, and eventually presents the results and discusses their implications (Dasigi et al., 2017). Prior work, as shown in Table 1, has proposed various schemes to analyze the structures of scientific articles. Parsing all the CORD-19 papers automatically and representing their structures using a semantic scheme (*e.g.*, background, method, result, etc.) will make it easier

---

[1]COVID-19 Research Aspect Dataset (CODA-19): http://CODA-19.org

	Corpus	Document type	Instance type	# of documents	# of sentence	Annotator type	# of classes	Public?
This work	**CORD-19**	**abstract**	**clause**	**10,966**	**103,978**	**crowd**	**5**	**yes**
Liakata et al.	ART†	paper	sentence	225	35,040	expert	18	yes
Ravenscroft et al.	MCCRA†	paper	sentence	50	8,501	expert	18	yes
Teufel and Moens	CONF	paper	sentence	80	12,188	expert	7	no
Kim et al.	MEDLINE	abstract	sentence	1,000	10,379	expert	6	no
Contractor et al.	PubMed	paper	sentence	50	8,569	expert	8	no
Morid et al.	UpToDate+PubMed	mixed	sentence	158	5,896	expert	8	no
Banerjee et al.	CONF+arXiv	paper	sentence	450	4,165	expert	3	no
Huang and Chen	NTHU database	abstract	sentence	597	3,394	expert	5	no
Agarwal and Yu	BioMed Central	paper	sentence	148	2,960	expert	5	no
Hara and Matsumoto	MEDLINE	abstract	sentence	200	2,390	expert	5	no
Zhao et al.	JOUR	mixed	sentence	...	2,000	expert	5	no
McKnight and Srinivasan	MEDLINE	abstract	sentence	204	1,532	...	4	no
Chung	PubMed	abstract	sentence	318	829	expert	4	no
Wu et al.	CiteSeer	abstract	sentence	106	709	expert	5	no
Dasigi et al.	PubMed	section	clause	75	<4,497*	expert	7	no
Ruch et al.	PubMed	abstract	sentence	100	...	...	4	no
Lin et al.	PubMed	abstract	sentence	49	...	expert	4	no

Table 1: Comparison of datasets, excluding structured abstracts. Abbreviations CONF and JOUR mean conference and journal papers, respectively. The dagger symbol † means that the corpus is self-curated. The "mixed" document type means that the dataset contains instances from both full papers and abstracts. Symbol "..." means unknown. The last "Public" column means if the dataset is publicly downloadable on the Internet. The symbol * means that the work only provides the number of clauses (sentence fragments).

for both humans and machines to comprehend and process the potentially life-saving information in these 13,000+ papers.

To reach good performance levels, modern automated language-understanding approaches often require large-scale human annotations as training data. Researchers traditionally relied on experts to annotate the structures of scientific papers, as shown in Table 1. However, producing such annotations for thousands of papers will be a prolonged process if we only employ experts, whose availability is much more limited than that of non-expert annotators. As a consequence, most of the human-annotated corpora that labeled the structures of scientific articles covered no more than one thousand papers (Table 1). Waiting to obtain expert annotations is too slow to respond to COVID-19, so we explored an alternative approach: using **non-expert crowds**, such as workers on Amazon Mechanical Turk (MTurk), to produce high-quality, useful annotations for thousands of scientific papers.

Researchers have used non-expert crowds to annotate text, for example, for machine translation (Wijaya et al., 2017; Gao et al., 2015; Yan et al., 2014; Zaidan and Callison-Burch, 2011; Post et al., 2012), natural language inference (Bowman et al., 2015; Khot et al., 2018), and medical report analysis (Maclean and Heer, 2013; Zhai et al., 2013; Good et al., 2014; Li et al., 2016).

While domain experts are still valuable in creating high-quality labels (Stubbs, 2013; Pustejovsky and Stubbs, 2012), and there are concerns about the use of MTurk (Fort et al., 2011; Cohen et al., 2016), employing non-expert crowds has been shown to be an effective and scalable approach to creating datasets. However, annotating *papers* is still often viewed as an expert task. The majority of the datasets only used experts (Table 1), or information provided by the paper's authors (Table 2), to denote the structures of scientific papers. One exception was the SOLVENT project by Chan *et al.* (2018). They recruited MTurk workers to annotate tokens (words) in paper abstracts with the research aspects (*e.g.*, Background, Mechanism, Finding). However, while the professional editors recruited from Upwork performed well, the MTurk workers' token-level accuracy was only 59%, which was insufficient for training good machine-learning models.

This paper introduces **CODA-19**, the <u>**COVID-19 Research Aspect Dataset**</u> and presents the first outcome of our exploration in using non-expert crowds for large-scale scholarly article annotation. CODA-19 contains 10,966 abstracts randomly selected from CORD-19. Each abstract was segmented into sentences, which were further divided into one or more shorter text fragments. All 168,286 text fragments in CODA-19 were labeled with a "research aspect," *i.e.*, **Background, Pur-**

	Corpus	Document	Instance	# of documents	# of sentence	# of classes	Public?
Dernoncourt and Lee	PubMed	structured abstract	sentence	195K	2.2M	5	yes
Jin and Szolovits	PubMed	structured abstract	sentence	24K	319K	7	yes
Huang et al.	MEDLINE	structured abstract	sentence	19K	526K	3	no
Boudin et al.	PubMed	structured abstract	sentence	260K	349K	3	no
Banerjee et al.	PubMed	structured abstract	sentence	20K	216K	3	no
Chung	PubMed	structured abstract	sentence	13K	156K	4	no
Shimbo et al.	MEDLINE	structured abstract	sentence	11K	114K	5	no
McKnight and Srinivasan	MEDLINE	structured abstract	sentence	7K	90K	4	no
Chung and Coiera	MEDLINE	structured abstract	sentence	3K	45K	5	no
Hirohata et al.	MEDLINE	structured abstract	sentence	683K	...	4	no
Lin et al.	MEDLINE	structured abstract	sentence	308K	...	4	no
Ruch et al.	PubMed	structured abstract	sentence	12K	...	4	no

Table 2: Comparison of datasets, leveraging structured abstracts that do not require human-labeling effort. Some works that involve human-labeled data are also present in Table 1.

pose, Method, Finding/Contribution, or Other. This annotation scheme was adapted from SOL-VENT (Chan et al., 2018), with minor changes. Figure 1 shows an example annotated abstract.

In our project, 248 crowd workers from MTurk were recruited and annotated the whole CODA-19 **within ten days**.[2] Nine different workers annotated each abstract. We aggregated the crowd labels for each text segment using majority voting.

The resulting crowd labels had a label accuracy of 82% when compared against the expert's labels of 129 abstracts. The inter-annotator agreement (Cohen's kappa) was 0.741 between the crowd labels and the expert labels, and 0.788 between two experts. We also established several classification baselines, showing the feasibility of automating such annotation tasks.

2 Related Work

A significant body of prior work has explored revealing or parsing the structures of scientific articles, including composing structured abstracts (Hartley, 2004), identifying argumentative zones (Teufel et al., 1999; Mizuta et al., 2006; Liakata et al., 2010), analyzing scientific discourse (de Waard and Maat, 2012; Dasigi et al., 2017; Banerjee et al., 2020), supporting paper writing (Wang et al., 2019; Huang and Chen, 2017), and representing papers to reduce information overload (de Waard et al., 2009). In this section, we review the datasets that were created to study the structures of scientific articles.

We sorted all the datasets that denoted the structures of scientific papers into two categories: *(i)*

datasets that used human labor to manually annotate the sentences in scientific articles (Table 1), and *(ii)* the datasets that leveraged the structured abstracts (Table 2).

In the first category, researchers recruited a group of annotators– often experts, such as medical doctors, biologists, or computer scientists– to manually label the sentences in papers with their research aspects (*e.g.*, Background, Method, Findings). CODA-19 belongs to the first category. Table 1 reviews the existing datasets in this category. To the best of our knowledge, only two other datasets can be downloaded from the Internet besides our work. Nearly all of the datasets of this kind were annotated by domain experts or researchers, which limited their size significantly. In Table 1, CODA-19 is the only dataset that contains more than 1,000 papers. Our work presents a scalable and efficient solution that employs non-expert crowd workers to annotate scientific papers. Furthermore, our labels were based on clauses (also referred to as sentence fragments or sub-sentences), which provide more detailed information than the majority of other works that used sentence-level annotations.

In the second category, researchers used the section titles that came with structured abstracts in scientific databases (*e.g.*, PubMed) to label sentences. A structured abstract is an abstract with distinctly labeled sections (*e.g.*, Introduction, Methods, Results) (Hartley, 2004). Different journals have different guidelines for section titles. To form a coherent and standardized dataset, researchers often mapped these different titles into a smaller set of labels. The sizes of the datasets in the second category (Table 2) were typically larger, because

---

[2]From April 19, 2020 to April 29, 2020, including worker training and a post-task survey.

they did not require extra annotating effort. This line of research is inspiring; however, assigning the same label to all the sentences in the same section overlooks the information granularity at the sentence level. Furthermore, not every journal uses the format of structured abstracts. The language used for describing a research work with a coherent paragraph might differ from the language used for presenting the work with a set of predetermined sections. Our work creates an in-domain dataset with high-quality labels for each sentence fragment in the 10,000+ abstracts, regardless of their formats.

3 CODA-19 Dataset Construction

CODA-19 has 10,966 abstracts that contain a total of 2,703,174 tokens and 103,978 sentences, which were divided into 168,286 segments. The data is released as an 80/10/10 training/validation/test split.

3.1 Annotation Scheme

CODA-19 uses a five-class annotation scheme to denote research aspects in scientific articles: **Background, Purpose, Method, Finding/Contribution, or Other**. Table 3 shows the full annotation guidelines we developed to instruct workers. We updated and expanded this guideline daily during the annotation process to address workers' questions and feedback. This scheme was adapted from SOLVENT (Chan et al., 2018), with three changes. First, we added an "Other" category. Articles in CORD-19 are broad and diverse (Colavizza et al., 2020), so it is unrealistic to annotate all of them into only four categories. We are also aware that CORD-19's data has occasional formatting or segmenting errors. These cases were to be put into the "Other" category. Second, we replaced the "Mechanism" category with "Method." Chan *et al.* created SOLVENT with the aim of discovering the analogies between research papers at scale. Our goal was to better understand the contribution of each paper, so we decided to use a more general word, "Method," to include research methods and procedures that cannot be characterized as "Mechanisms." Also, biomedical literature uses the word "mechanism" widely, which could also be confusing to workers. Third, we modified the name "Finding" to "Finding/Contribution" to allow broader contributions that are not usually viewed as "findings."

3.2 Data Preparation

We used Stanford CoreNLP (Manning et al., 2014) to tokenize and segment sentences for all the abstracts in CORD-19. We further used commas (,), semicolons (;), and periods (.) to split each sentence into shorter fragments, where a fragment has no fewer than six tokens (including punctuation marks) and no orphan parentheses.

As of April 15, 2020, 29,306 articles in CORD-19 had a non-empty abstract. The average abstract had 9.73 sentences (SD = 8.44), which were further divided into 15.75 text segments (SD = 13.26). Each abstract had 252.36 tokens (SD = 192.89) on average. We filtered out 538 (1.84%) abstracts with only one sentence because many of them had formatting errors. We also removed 145 (0.49%) abstracts that had more than 1,200 tokens to keep the working time for each task under five minutes (see Section 3.4). We randomly selected 11,000 abstracts from the remaining data for annotation. During the annotation process, workers informed us that a few articles were not in English. We identified these automatically using `langdetect`[3] and excluded them.

3.3 Interface Design

Figure 2 shows the worker interface we designed to guide workers to read and label all the text segments in an abstract. The interface showed the instruction on the top (Figure 2a) and presented the task in three steps. In Step 1, the worker was instructed to spend ten seconds quickly glancing at the abstract. The goal was to get a high-level sense of the topic rather than to fully understand the abstract. In Step 2, we showed the main annotation interface (Figure 2b), where the worker worked through each text segment and selected the most appropriate category for each segment one by one. In Step 3, the worker reviewed the labeled text segments (Figure 2c) and went back to Step 2 to fix any problems.

3.4 Annotation Procedure

Worker Training and Recruitment We first created a qualification Human Intelligence Task (HIT) to recruit workers on MTurk ($1/HIT). The workers needed to watch a five-minute video to learn the scheme, go through an interactive tutorial to learn the interface, and sign a consent form to

---

[3]langdetect: https://github.com/Mimino666/langdetect

Aspect	Annotation Guideline
Background	"Background" text segments answer one or more of these questions: • Why is this problem important? • What relevant works have been created before? • What is still missing in the previous works? • What are the high-level research questions? • How might this help other research or researchers?
Purpose	"Purpose" text segments answer one or more of these questions: • What specific things do the researchers want to do? • What specific knowledge do the researchers want to gain? • What specific hypothesis do the researchers want to test?
Method	"Method" text segments answer one or more of these questions: • How did the researchers do the work or find what they sought? • What are the procedures and steps of the research?
Finding/ Contribution	"Finding/Contribution" text segments answer one or more of these questions: • What did the researchers find out? • Did the proposed methods work? • Did the thing behave as the researchers expected?
Other	• Text segments that do *not* fit into any of the four categories above. • Text segments that are *not* part of the article. • Text segments that are *not* in English. • Text segments that contain *only* reference marks (*e.g.*, "[1,2,3,4,5") or dates (*e.g.*, "April 20, 2008"). • Captions for figures and tables (*e.g.* "Figure 1: Experimental Result of ...") • Formatting errors. • Text segments the annotator does not know or is not sure about.

Table 3: CODA-19's annotation guideline for crowd workers.

obtain the qualification. We granted custom qualifications to 400 workers who accomplished the qualification HIT. Only the workers with this qualification could do our tasks.[4]

Posting Tasks in Smaller Batches We divided the 11,000 abstracts into smaller batches, where each batch had no more than 1,000 abstracts. Each abstract forms a single HIT. We recruited nine different workers through nine assignments to label each abstract. Our strategy was to post one batch at a time. When a batch was finished, we assessed its data quality, sent feedback to workers, and blocked workers who showed consistently low accuracy before proceeding with the next batch.

Worker Wage and Total Cost We aimed to pay an hourly wage of $10. The working time of an abstract was estimated by the average reading speed of English native speakers, *i.e.*, 200-300 words per minute (Siegenthaler et al., 2012). For an abstract, we rounded up ($\#token/250$) to an integer as the estimated working time in minutes, and paid ($\$0.05 +$ Estimated Working Minutes $\times \$0.17$) for it. As a result, 59.49% of our HITs were priced at $0.22, 36.41% were at $0.39, 2.74% were at

$0.56, 0.81% were at $0.73, and 0.55% were at $0.90. We posted nine assignments per HIT. With the 20% MTurk fee, coding each abstract (using nine workers) cost $3.21 on average.

In this project, each worker received an average of ($\$3.21/9)/1.2 = \$0.297$ for annotating one abstract. We empirically learned that the CS Expert (see Section 4) spent an average of 50.8 seconds (SD=10.4, N=10) to annotate an abstract, yielding an estimated hourly wage of $\$0.297 \times (60 \times 60/50.8) = \$21.05$; and the MTurk workers in SOLVENT took a median of 1.3 minutes to annotate one abstract (Chan et al., 2018), yielding an estimated hourly wage of $\$0.297 \times (60/1.3) = \$13.71$. We thus believe that the actual hourly wage for our workers was close to or over $10.

3.5 Label Aggregation

The final labels in CODA-19 were acquired by majority vote for crowd labels (excluding the labels from blocked workers). For each batch of HITs, we manually examined the labels from workers who frequently disagreed with the majority-voted labels (Section 3.4). If a worker had abnormally low accuracy or was apparently spamming, we retracted the worker's qualification to prevent him/her from taking future tasks. We excluded the labels from these removed workers when aggregating the final

---

[4]Four built-in MTurk qualifications were also used: Locale (US Only), HIT Approval Rate ($\geq$98%), Number of Approved HITs ($\geq$3000), and the Adult Content Qualification.

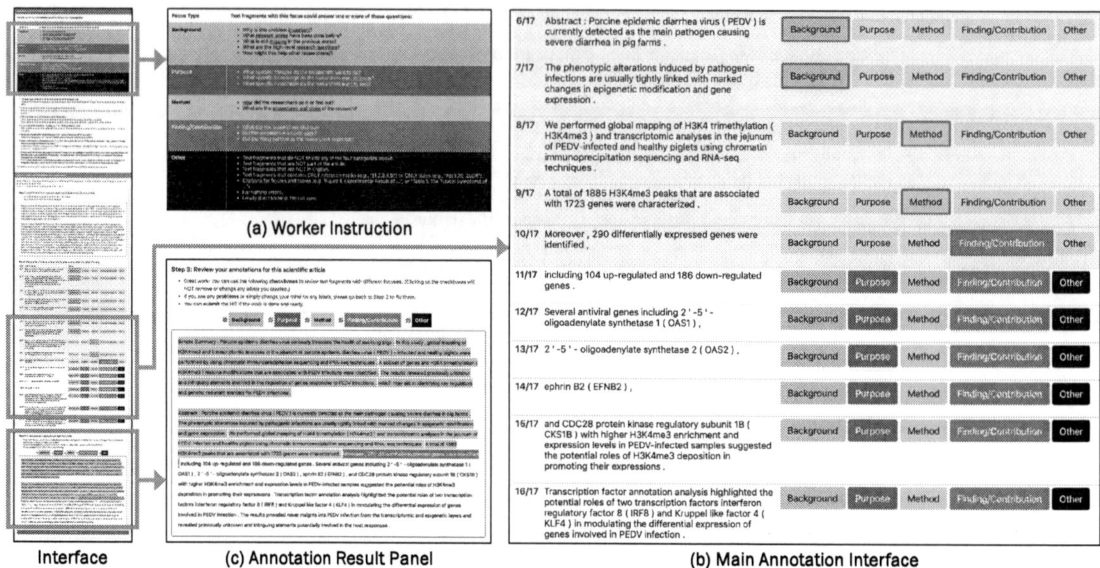

Figure 2: The worker interface used to construct CODA-19.

labels. Note that there can be ties when two or more aspects received the same highest number of votes (*e.g.,* 4/4/1 or 3/3/3). We resolved ties by using the following tiebreakers in order: Finding, Method, Purpose, Background, Other.

4 Data Quality Assessment

We worked with a biomedical expert and a computer scientist to assess label quality; both experts are co-authors of this paper. The biomedical expert (the "Bio" Expert in Table 4) is an M.D. and a Ph.D. in Genetics and Genomics. She is now a resident physician in pathology at the University of California, San Francisco. The other expert (the "CS" Expert in Table 4) has a Ph.D. in Computer Science and is currently a Project Scientist at Carnegie Mellon University.

Both experts annotated the same 129 abstracts randomly selected from CODA-19. The experts used the same interface as that of the workers (Figure 2). We used scikit-learn's implementation (Pedregosa et al., 2011) to compute the inter-annotator agreement (Cohen's kappa). The kappa between the two experts was 0.788. Table 4 shows the aggregated crowd label's accuracy, along with the precision, recall, and F1-score of each research aspect. CODA-19's labels have an accuracy of 0.82 and a kappa of 0.74 when compared against the two experts' labels. It is noteworthy that when we compared labels between the two experts, the accuracy (0.850) and kappa (0.788) were only slightly

higher. The crowd workers performed best in labeling Background and Finding, and they had nearly perfect precision for the Other category. Figure 3 shows the normalized confusion matrix for the aggregated crowd labels versus the biomedical expert's labels. Many Purpose segments were mislabeled as Background, which might indicate more ambiguous cases between these two categories. During the annotation period, we received several emails from workers asking about the distinctions between these two aspects. For example, do "potential applications of the proposed work" count as Background or Purpose?

5 Classification Baselines

We further examined the capacity of machines for annotating research aspects automatically. Seven baseline models were implemented: Linear SVM, Random Forest, Multinomial Naive Bayes (MNB), CNN, LSTM, BERT, and SciBERT.

Data Preprocessing The tf-idf feature was used for Linear SVM and Random Forest. We turned all words into lowercase and removed those with frequency lower than five. The final tf-idf feature contained 16,775 dimensions. Two variations of feature were used for MNB: the n-gram counts feature and the n-gram tf-idf feature. Using a grid search method, the n-gram counts feature, combining unigram, bigram, and trigram with a minimum frequency of three, yielded the best result. The final n-gram feature contained 181,391 dimensions.

Eval. Label	Gold Label	Background			Purpose			Method			Finding			Other			acc	kappa
		P	R	F1	P	R	F1	P	R	F1	P	R	F1	P	R	F1		
Crowd	Bio	.827	.911	.867	.427	.662	.519	.783	.710	.744	.874	.838	.856	.986	.609	.753	.822	.741
Crowd	CS	.846	.883	.864	.700	.611	.653	.818	.633	.714	.800	.931	.860	.986	.619	.761	.821	.745
CS	Bio	.915	.966	.940	.421	.746	.538	.670	.785	.723	.958	.789	.865	.867	.852	.860	.850	.788

Table 4: Crowd performance using both Bio Expert and CS Expert as the gold standard. CODA-19's labels have an accuracy of 0.82 and a kappa of 0.74, when compared against two experts' labels. It is noteworthy that when we compared labels between two experts, the accuracy (0.850) and kappa (0.788) were only slightly higher.

Model	Background			Purpose			Method			Finding			Other			Accuracy
	P	R	F1	P	R	F1	P	R	F1	P	R	F1	P	R	F1	
# Sample	5062			821			2140			6890			562			15475
SVM	.658	.703	.680	.621	.446	.519	.615	.495	.549	.697	.729	.712	.729	.699	.714	.672
RF	.671	.632	.651	.696	.365	.479	**.716**	.350	.471	.630	**.787**	.699	.674	.742	.706	.652
MNB-count	.654	.714	.683	.549	.514	.531	.570	.585	.577	.711	.691	.701	**.824**	.425	.561	.665
MNB-tfidf	.655	.683	.669	.673	391	.495	.640	.469	.541	.661	.754	.704	.757	.383	.508	.659
CNN	.649	.706	.676	.612	.512	.557	.596	.562	.579	.726	.702	.714	.743	.795	.768	.677
LSTM	.655	.706	.680	**.700**	.464	.558	.634	.508	.564	.700	.724	.711	.682	.770	.723	.676
BERT	.719	.759	.738	.585	**.639**	.611	.680	.612	.644	.777	.752	.764	.773	**.874**	.820	.733
SciBERT	**.733**	**.768**	**.750**	.616	.636	**.626**	.715	**.636**	**.673**	**.783**	.775	**.779**	.794	.852	**.822**	**.749**

Table 5: Baseline performance of automatic labeling using the crowd labels of CODA-19. SciBERT achieves highest accuracy of 0.749 and outperforms other models in every aspects.

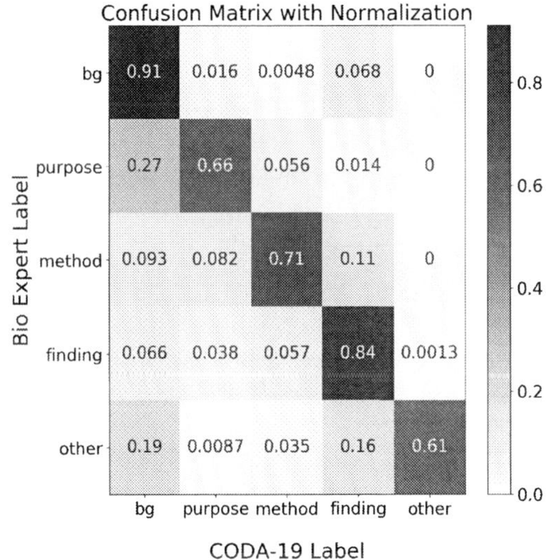

Figure 3: The normalized confusion matrix for the CODA-19 labels versus the biomedical expert's labels.

The n-gram tf-idf feature combining unigram, bigram, and trigram with minimum frequency of 10 yielded the best result where the dimensions were 41,966. For deep-learning approaches, the vocabulary size was 16,135, where tokens with a frequency lower than five were replaced by <UNK>. Sequences were padded with <PAD> if they contained less than 60 tokens, and were truncated if they contained more than 60 tokens.

Models Machine-learning approaches were implemented using Scikit-learn (Pedregosa et al., 2011). Deep-learning approaches were implemented using PyTorch (Paszke et al., 2019). The following are the training setups.

- **Linear SVM:** We did a grid search for hyper-parameters and found that $C = 1$, $tol = 0.001$, and *hinge loss* yielded the best results.

- **Random Forest:** With the grid search, 150 estimators yielded the best result.

- **Multinomial Naive Bayes (MNB):** Several important early works used Naive Bayes models for text classification (Rennie et al., 2003; McCallum et al., 1998). When using n-gram counts as the feature, the default parameter, $alpha = 1.0$, yielded the best result. For the one using n-gram tf-idf feature, $alpha = 0.5$ yielded the best result.

- **CNN:** The classic CNN (Kim, 2014) was implemented. Three kernel sizes (3, 4, 5) were used, each with 100 filters. The word embedding size was 256. A dropout rate of 0.3 and L2 regularization with weight $1e - 6$ were used when training. We used the Adam optimizer with a learning rate of $5e - 5$. The model was trained for 50 epochs, and the one with the highest validation score was kept for testing.

- **LSTM:** We used 10 LSTM layers to encode the sequence. The encoded vector was then passed through a dense layer for classification. The word embedding size and the LSTM hidden size were both 256. The rest of the hyperparameter and training settings were the same as that of the CNN model.

- **BERT:** Hugging Face's implementation (Wolf et al., 2019) of the Pretrained BERT (Devlin et al., 2018) was used for fine-tuning. We fine-tuned the pretrained model with a learning rate of $3e - 7$ for 50 epochs. Early stopping was used when no improvement occurred in the validation accuracy for five consecutive epochs. The model with the highest validation score was kept for testing.

- **SciBERT:** Hugging Face's implementation (Wolf et al., 2019) of the Pretrained SciBERT (Beltagy et al., 2019) was used for finetuning. The fine-tuning setting is the same as that of the BERT model.

Result Table 5 shows the results for the six baseline models: SciBERT preformed the best in overall accuracy. When looking at each aspect, all the models performed well in classifying Background, Finding, and Other, while identifying Purpose and Method was more challenging.

6 Discussion

Annotating scientific papers was often viewed as an expert task that was difficult or impossible for non-expert annotators to do. Many datasets that labeled scientific papers were thus produced by small groups of experts. For example, two researchers manually created the ACL RD-TEC 2.0, a dataset that contains 300 scientific abstracts (QasemiZadeh and Schumann, 2016). A group of annotators "with rich experience in biomedical content curation" created MedMentions, a corpus containing 4,000 abstracts (Mohan and Li, 2019). Many datasets used in biomedical NLP shared tasks were manually created by the organizers and/or their students, such as ScienceIE in SemEval'17 (Augenstein et al., 2017) and Relation Extraction in SemEval'18 (Gábor et al., 2018). Our work suggests that non-expert

crowds, such as crowd workers or volunteers, can be used for these types of data-labeling tasks.

Meanwhile, some prior work used crowd workers to annotate pieces of lower-level information on papers or medical documents, such as images (Heim et al., 2018), named entities (*e.g.,*, medical terms (Mohan and Li, 2019), disease (Good et al., 2014), and medicine (Abaho et al., 2019).) Our work shows that to a certain extent, crowd workers can comprehend the high-level structures and discourses in papers, and therefore can be assigned more complex, higher-level tasks.

7 Conclusion and Future Work

This paper introduces CODA-19, a humanannotated dataset that codes the Background, Purpose, Method, Finding/Contribution, and Other sections of 10,966 English abstracts in the COVID-19 Open Research Dataset. CODA-19 was created by a group of MTurk workers, who achieved labeling quality comparable to that of experts. We demonstrated that a non-expert crowd can be rapidly employed at scale to join the fight against COVID-19.

One future direction is to improve classification performance. We evaluated the automatic labels against the biomedical expert's labels, and the SciBERT model achieved an accuracy of 0.774 and a Cohen's kappa of 0.667, indicating potential for further improvement. Furthermore, one motivation to automatically annotate research aspects is to help search and information extraction (Teufel et al., 1999). We have teamed up with the group who created COVIDSeer[5] to explore the possible uses of CODA-19 in such systems.

Acknowledgments

This project was supported by the Huck Institutes of the Life Sciences' Coronavirus Research Seed Fund (CRSF) and the College of IST COVID-19 Seed Fund at Penn State University. We thank the crowd workers for participating in this project and providing useful feedback. We thank VoiceBunny Inc. for granting a 60% discount for the voiceover for the worker tutorial video in support of projects relevant to COVID-19. We also thank Tiffany Knearem, Shih-Hong (Alan) Huang, Joseph Chee Chang, and Frank Ritter for great discussion and useful feedback.

---

[5]CovidSeer: https://covidseer.ist.psu.edu/

References

Micheal Abaho, Danushka Bollegala, Paula Williamson, and Susanna Dodd. 2019. Correcting crowdsourced annotations to improve detection of outcome types in evidence based medicine. In *CEUR Workshop Proceedings*, volume 2429, pages 1–5.

Shashank Agarwal and Hong Yu. 2009. Automatically classifying sentences in full-text biomedical articles into introduction, methods, results and discussion. *Bioinformatics*, 25(23):3174–3180.

Allen Institute For AI. 2020. Covid-19 open research dataset challenge (cord-19).

Michael Alley. 1996. The craft of scientific writing. Technical report, Springer.

Isabelle Augenstein, Mrinal Das, Sebastian Riedel, Lakshmi Vikraman, and Andrew McCallum. 2017. Semeval 2017 task 10: Scienceie-extracting keyphrases and relations from scientific publications. *arXiv preprint arXiv:1704.02853*.

Soumya Banerjee, Debarshi Kumar Sanyal, Samiran Chattopadhyay, Plaban Kumar Bhowmick, and Parthapratim Das. 2020. Segmenting scientific abstracts into discourse categories: A deep learning-based approach for sparse labeled data. *arXiv preprint arXiv:2005.05414*.

Iz Beltagy, Kyle Lo, and Arman Cohan. 2019. Scibert: A pretrained language model for scientific text. In *Proceedings of the 2019 Conference on Empirical Methods in Natural Language Processing and the 9th International Joint Conference on Natural Language Processing (EMNLP-IJCNLP)*, pages 3606–3611.

Florian Boudin, Jian-Yun Nie, Joan C Bartlett, Roland Grad, Pierre Pluye, and Martin Dawes. 2010. Combining classifiers for robust pico element detection. *BMC medical informatics and decision making*, 10(1):29.

Samuel R. Bowman, Gabor Angeli, Christopher Potts, and Christopher D. Manning. 2015. A large annotated corpus for learning natural language inference. In *Proceedings of the 2015 Conference on Empirical Methods in Natural Language Processing*, pages 632–642, Lisbon, Portugal. Association for Computational Linguistics.

Joel Chan, Joseph Chee Chang, Tom Hope, Dafna Shahaf, and Aniket Kittur. 2018. Solvent: A mixed initiative system for finding analogies between research papers. *Proceedings of the ACM on Human-Computer Interaction*, 2(CSCW):1–21.

Grace Chung and Enrico Coiera. 2007. A study of structured clinical abstracts and the semantic classification of sentences. In *Biological, translational, and clinical language processing*, pages 121–128.

Grace Y Chung. 2009. Sentence retrieval for abstracts of randomized controlled trials. *BMC medical informatics and decision making*, 9(1):10.

K Bretonnel Cohen, Karën Fort, Gilles Adda, Sophia Zhou, and Dimeji Farri. 2016. Ethical issues in corpus linguistics and annotation: Pay per hit does not affect effective hourly rate for linguistic resource development on amazon mechanical turk. In *LREC... International Conference on Language Resources & Evaluation:[proceedings]. International Conference on Language Resources and Evaluation*, volume 2016, page 8. NIH Public Access.

Giovanni Colavizza, Rodrigo Costas, Vincent A. Traag, Nees Jan van Eck, Thed van Leeuwen, and Ludo Waltman. 2020. A scientometric overview of cord-19. *bioRxiv*.

Danish Contractor, Yufan Guo, and Anna Korhonen. 2012. Using argumentative zones for extractive summarization of scientific articles. In *Proceedings of COLING 2012*, pages 663–678.

Pradeep Dasigi, Gully APC Burns, Eduard Hovy, and Anita de Waard. 2017. Experiment segmentation in scientific discourse as clause-level structured prediction using recurrent neural networks. *arXiv preprint arXiv:1702.05398*.

Franck Dernoncourt and Ji Young Lee. 2017. Pubmed 200k rct: a dataset for sequential sentence classification in medical abstracts. *arXiv preprint arXiv:1710.06071*.

Jacob Devlin, Ming-Wei Chang, Kenton Lee, and Kristina Toutanova. 2018. Bert: Pre-training of deep bidirectional transformers for language understanding. *arXiv preprint arXiv:1810.04805*.

Karën Fort, Gilles Adda, and K Bretonnel Cohen. 2011. Amazon mechanical turk: Gold mine or coal mine? *Computational Linguistics*, 37(2):413–420.

Kata Gábor, Davide Buscaldi, Anne-Kathrin Schumann, Behrang QasemiZadeh, Haifa Zargayouna, and Thierry Charnois. 2018. Semeval-2018 task 7: Semantic relation extraction and classification in scientific papers. In *Proceedings of The 12th International Workshop on Semantic Evaluation*, pages 679–688.

Mingkun Gao, Wei Xu, and Chris Callison-Burch. 2015. Cost optimization in crowdsourcing translation. In *Proceedings of the 2015 Conference of the North American Chapter of the Association for Computational Linguistics (NAACL 2015)*, Denver, Colorado.

Benjamin M Good, Max Nanis, Chunlei Wu, and Andrew I Su. 2014. Microtask crowdsourcing for disease mention annotation in pubmed abstracts. In *Pacific Symposium on Biocomputing Co-Chairs*, pages 282–293. World Scientific.

Kazuo Hara and Yuji Matsumoto. 2007. Extracting clinical trial design information from medline abstracts. *New Generation Computing*, 25(3):263–275.

James Hartley. 2004. Current findings from research on structured abstracts. *Journal of the Medical Library Association*, 92(3):368.

Eric Heim, Tobias Roß, Alexander Seitel, Keno März, Bram Stieltjes, Matthias Eisenmann, Johannes Lebert, Jasmin Metzger, Gregor Sommer, Alexander W Sauter, et al. 2018. Large-scale medical image annotation with crowd-powered algorithms. *Journal of Medical Imaging*, 5(3):034002.

Kenji Hirohata, Naoaki Okazaki, Sophia Ananiadou, and Mitsuru Ishizuka. 2008. Identifying sections in scientific abstracts using conditional random fields. In *Proceedings of the Third International Joint Conference on Natural Language Processing: Volume-I*.

Hen-Hsen Huang and Hsin-Hsi Chen. 2017. Disa: A scientific writing advisor with deep information structure analysis. In *IJCAI*, pages 5229–5231.

Ke-Chun Huang, I-Jen Chiang, Furen Xiao, Chun-Chih Liao, Charles Chih-Ho Liu, and Jau-Min Wong. 2013. Pico element detection in medical text without metadata: Are first sentences enough? *Journal of biomedical informatics*, 46(5):940–946.

Natalia B Hubbs, Mareena M Whisby-Pitts, and Jonathan L McMurry. 2019. Kinetic analysis of bacteriophage sf6 binding to outer membrane protein a using whole virions. *bioRxiv*, page 509141.

Di Jin and Peter Szolovits. 2018. Pico element detection in medical text via long short-term memory neural networks. In *Proceedings of the BioNLP 2018 workshop*, pages 67–75.

Tushar Khot, Ashish Sabharwal, and Peter Clark. 2018. SciTail: A textual entailment dataset from science question answering. In *AAAI*.

Su Nam Kim, David Martinez, Lawrence Cavedon, and Lars Yencken. 2011. Automatic classification of sentences to support evidence based medicine. In *BMC bioinformatics*, volume 12, page S5. Springer.

Yoon Kim. 2014. Convolutional neural networks for sentence classification. *arXiv preprint arXiv:1408.5882*.

Tong Li, Àlex Bravo, Laura I Furlong, Benjamin Good, and Andrew Su. 2016. A crowdsourcing workflow for extracting chemical-induced disease relations from free text. *Database*, 2016:baw051.

Maria Liakata, Larisa N Soldatova, et al. 2009. Semantic annotation of papers: Interface & enrichment tool (sapient). In *Proceedings of the Workshop on Current Trends in Biomedical Natural Language Processing*, pages 193–200. Association for Computational Linguistics.

Maria Liakata, Simone Teufel, Advaith Siddharthan, and Colin Batchelor. 2010. Corpora for the conceptualisation and zoning of scientific papers. In *Proceedings of the Seventh conference on International Language Resources and Evaluation (LREC'10)*.

Jimmy Lin, Damianos Karakos, Dina Demner-Fushman, and Sanjeev Khudanpur. 2006. Generative content models for structural analysis of medical abstracts. In *Proceedings of the hlt-naacl bionlp workshop on linking natural language and biology*, pages 65–72.

Diana Maclean and Jeffrey Heer. 2013. Identifying medical terms in patient-authored text: A crowdsourcing-based approach. *Journal of the American Medical Informatics Association : JAMIA*, 20.

Christopher D. Manning, Mihai Surdeanu, John Bauer, Jenny Finkel, Steven J. Bethard, and David McClosky. 2014. The Stanford CoreNLP natural language processing toolkit. In *Association for Computational Linguistics (ACL) System Demonstrations*, pages 55–60.

Andrew McCallum, Kamal Nigam, et al. 1998. A comparison of event models for naive bayes text classification. In *AAAI-98 workshop on learning for text categorization*, volume 752, pages 41–48. Citeseer.

Larry McKnight and Padmini Srinivasan. 2003. Categorization of sentence types in medical abstracts. In *AMIA Annual Symposium Proceedings*, volume 2003, page 440. American Medical Informatics Association.

Yoko Mizuta, Anna Korhonen, Tony Mullen, and Nigel Collier. 2006. Zone analysis in biology articles as a basis for information extraction. *International journal of medical informatics*, 75(6):468–487.

Sunil Mohan and Donghui Li. 2019. Medmentions: a large biomedical corpus annotated with umls concepts. *arXiv preprint arXiv:1902.09476*.

Mohammad Amin Morid, Marcelo Fiszman, Kalpana Raja, Siddhartha R Jonnalagadda, and Guilherme Del Fiol. 2016. Classification of clinically useful sentences in clinical evidence resources. *Journal of biomedical informatics*, 60:14–22.

Adam Paszke, Sam Gross, Francisco Massa, Adam Lerer, James Bradbury, Gregory Chanan, Trevor Killeen, Zeming Lin, Natalia Gimelshein, Luca Antiga, Alban Desmaison, Andreas Kopf, Edward Yang, Zachary DeVito, Martin Raison, Alykhan Tejani, Sasank Chilamkurthy, Benoit Steiner, Lu Fang, Junjie Bai, and Soumith Chintala. 2019. Pytorch: An imperative style, high-performance deep learning library. In H. Wallach, H. Larochelle, A. Beygelzimer, F. d'Alché-Buc, E. Fox, and R. Garnett, editors, *Advances in Neural Information Processing Systems 32*, pages 8024–8035. Curran Associates, Inc.

Fabian Pedregosa, Gaël Varoquaux, Alexandre Gramfort, Vincent Michel, Bertrand Thirion, Olivier Grisel, Mathieu Blondel, Peter Prettenhofer, Ron Weiss, Vincent Dubourg, Jake Vanderplas, Alexandre Passos, David Cournapeau, Matthieu Brucher, Matthieu Perrot, and Édouard Duchesnay. 2011. Scikit-learn: Machine learning in python. *J. Mach. Learn. Res.*, 12(null):2825–2830.

Matt Post, Chris Callison-Burch, and Miles Osborne. 2012. Constructing parallel corpora for six Indian languages via crowdsourcing. In *Proceedings of the Seventh Workshop on Statistical Machine Translation*, pages 401–409, Montréal, Canada. Association for Computational Linguistics.

James Pustejovsky and Amber Stubbs. 2012. *Natural Language Annotation for Machine Learning: A guide to corpus-building for applications.* ” O'Reilly Media, Inc.”.

Behrang QasemiZadeh and Anne-Kathrin Schumann. 2016. The acl rd-tec 2.0: A language resource for evaluating term extraction and entity recognition methods. In *Proceedings of the Tenth International Conference on Language Resources and Evaluation (LREC'16)*, pages 1862–1868.

James Ravenscroft, Anika Oellrich, Shyamasree Saha, and Maria Liakata. 2016. Multi-label annotation in scientific articles-the multi-label cancer risk assessment corpus. In *Proceedings of the Tenth International Conference on Language Resources and Evaluation (LREC'16)*, pages 4115–4123.

Jason D Rennie, Lawrence Shih, Jaime Teevan, and David R Karger. 2003. Tackling the poor assumptions of naive bayes text classifiers. In *Proceedings of the 20th international conference on machine learning (ICML-03)*, pages 616–623.

Patrick Ruch, Celia Boyer, Christine Chichester, Imad Tbahriti, Antoine Geissbühler, Paul Fabry, Julien Gobeill, Violaine Pillet, Dietrich Rebholz-Schuhmann, Christian Lovis, et al. 2007. Using argumentation to extract key sentences from biomedical abstracts. *International journal of medical informatics*, 76(2-3):195–200.

Masashi Shimbo, Takahiro Yamasaki, and Yuji Matsumoto. 2003. Using sectioning information for text retrieval: a case study with the medline abstracts. In *Proceedings of Second International Workshop on Active Mining (AM'03)*.

Eva Siegenthaler, Yves Bochud, Per Bergamin, and Pascal Wurtz. 2012. Reading on lcd vs e-ink displays: effects on fatigue and visual strain. *Ophthalmic and Physiological Optics*, 32(5):367–374.

Amber C Stubbs. 2013. A methodology for using professional knowledge in corpus. *Waltham, MA: Brandeis University*.

Simone Teufel and Marc Moens. 2002. Summarizing scientific articles: experiments with relevance and rhetorical status. *Computational linguistics*, 28(4):409–445.

Simone Teufel et al. 1999. *Argumentative zoning: Information extraction from scientific text*. Ph.D. thesis, Citeseer.

A. de Waard, S. Buckingham Shum, A. Carusi, J. Park, M. Samwald, and Á. Sándor. 2009. Hypotheses, evidence and relationships: The hyper approach for representing scientific knowledge claims. In *Proceedings 8th International Semantic Web Conference, Workshop on Semantic Web Applications in Scientific Discourse. Lecture Notes in Computer Science, Springer Verlag: Berlin.*

Anita de Waard and Henk Pander Maat. 2012. Verb form indicates discourse segment type in biological research papers: Experimental evidence. *Journal of English for Academic Purposes*, 11(4):357–366.

Lucy Lu Wang, Kyle Lo, Yoganand Chandrasekhar, Russell Reas, Jiangjiang Yang, Darrin Eide, Kathryn Funk, Rodney Michael Kinney, Ziyang Liu, William. Merrill, Paul Mooney, Dewey A. Murdick, Devvret Rishi, Jerry Sheehan, Zhihong Shen, Brandon Stilson, Alex D. Wade, Kuansan Wang, Christopher Wilhelm, Boya Xie, Douglas M. Raymond, Daniel S. Weld, Oren Etzioni, and Sebastian Kohlmeier. 2020. Cord-19: The covid-19 open research dataset. *ArXiv*, abs/2004.10706.

Qingyun Wang, Lifu Huang, Zhiying Jiang, Kevin Knight, Heng Ji, Mohit Bansal, and Yi Luan. 2019. Paperrobot: Incremental draft generation of scientific ideas. In *Proceedings of the 57th Annual Meeting of the Association for Computational Linguistics*, pages 1980–1991.

Derry Wijaya, Brendan Callahan, John Hewitt, Jie Gao, Xiao Ling, Marianna Apidianaki, and Chris Callison-Burch. 2017. Learning translations via matrix completion. In *Conference on Empirical Methods in Natural Language Processing*, Copenhagen, Denmark.

Thomas Wolf, Lysandre Debut, Victor Sanh, Julien Chaumond, Clement Delangue, Anthony Moi, Pierric Cistac, Tim Rault, R'emi Louf, Morgan Funtowicz, and Jamie Brew. 2019. Huggingface's transformers: State-of-the-art natural language processing. *ArXiv, abs/1910.03771*

Jian-Chen Wu, Yu-Chia Chang, Hsien-Chin Liou, and Jason S Chang. 2006. Computational analysis of move structures in academic abstracts. In *Proceedings of the COLING/ACL 2006 Interactive Presentation Sessions*, pages 41–44.

Rui Yan, Mingkun Gao, Ellie Pavlick, and Chris Callison-Burch. 2014. Are two heads are better than one? crowdsourced translation via a two-step collaboration between translators and editors. In *The*

52nd Annual Meeting of the Association for Computational Linguistics, Baltimore, Maryland. Association for Computional Linguistics.

Omar F. Zaidan and Chris Callison-Burch. 2011. Crowdsourcing translation: Professional quality from non-professionals. In *Proceedings of the 49th Annual Meeting of the Association for Computational Linguistics: Human Language Technologies*, pages 1220–1229, Portland, Oregon, USA. Association for Computational Linguistics.

Haijun Zhai, Todd Lingren, Louise Deleger, Qi Li, Megan Kaiser, Laura Stoutenborough, and Imre Solti. 2013. Web 2.0-based crowdsourcing for high-quality gold standard development in clinical natural language processing. *Journal of medical Internet research*, 15:e73.

Jin Zhao, Praveen Bysani, and Min-Yen Kan. 2012. Exploiting classification correlations for the extraction of evidence-based practice information. In *AMIA Annual Symposium Proceedings*, volume 2012, page 1070. American Medical Informatics Association.

Information Retrieval and Extraction on COVID-19 Clinical Articles Using Graph Community Detection and Bio-BERT Embeddings

**Debasmita Das, Yatin Katyal, Janu Verma Shashank Dubey, Aakash Deep Singh,
Kushagra Agarwal, Sourojit Bhaduri, Rajesh Kumar Ranjan**
Mastercard AI Garage, Gurgaon, India
{firstname.secondname}@mastercard.com

Abstract

In this paper, we present an information retrieval system on a corpus of scientific articles related to COVID-19. We build a similarity network on the articles where similarity is determined via shared citations and biological domain-specific sentence embeddings. Ego-splitting community detection on the article network is employed to cluster the articles and then the queries are matched with the clusters. Extractive summarization using BERT and PageRank methods is used to provide responses to the query. We also provide a Question-Answer bot on a small set of intents to demonstrate the efficacy of our model for an information extraction module.

1 Methodology

We briefly describe our method here.

1. **Network of the articles:** We build a *citation graph* of the articles in the corpus where nodes corresponds to the papers and the edges are determined by the citations of the papers. If two papers have a common citation, then we add an edge between them. In addition, we include edges between papers if their semantic similarity is greater than a threshold. W use *BioSentVec* (Chen et al., 2019) which is trained on a corpus of about 30 million clinical and bio-medical research articles from the public databases - PubMed and MIMIC-III. *BioSentVec* provides *700-dimensional* sentence embeddings.

2. **Clustering of the articles:** We use **ego-splitting** (Epasto et al., 2017) based *community detection* algorithm to partition the articles in the network into clusters. The clusters are then studied qualitatively and we manually assign appropriate labels to the clusters.

3. **Mapping queries to the clusters:** A a given query, we find clusters that are closely related to the query. This mapping is facilitated by BioBERT embeddings (Lee et al., 2020) of the queries and the article titles. This helps in reducing the search space of the query to only within the most relevant clusters.

4. **Information Retrieval:** Focusing on the articles in the clusters relevant to the query, we use BioBERT embeddings of the whole articles (title, abstract, and body) to find the best matched articles within the clusters. Using a PageRank based procedure, we also extract best matching sentences to the query from within the most similar documents.

References

Qingyu Chen, Yifan Peng, and Zhiyong Lu. 2019. Biosentvec: creating sentence embeddings for biomedical texts. In *2019 IEEE International Conference on Healthcare Informatics (ICHI)*, pages 1–5. IEEE.

Alessandro Epasto, Silvio Lattanzi, and Renato Paes Leme. 2017. Ego-splitting framework: From non-overlapping to overlapping clusters. In *Proceedings of the 23rd ACM SIGKDD International Conference on Knowledge Discovery and Data Mining*, pages 145–154.

Jinhyuk Lee, Wonjin Yoon, Sungdong Kim, Donghyeon Kim, Sunkyu Kim, Chan Ho So, and Jaewoo Kang. 2020. Biobert: a pre-trained biomedical language representation model for biomedical text mining. *Bioinformatics*, 36(4):1234–1240.

What Are People Asking About COVID-19?
A Question Classification Dataset

Jerry Wei♦ **Chengyu Huang**♦ **Soroush Vosoughi**♥ **Jason Wei**♥

♦ProtagoLabs ♦International Monetary Fund ♥Dartmouth College

jerry.weng.wei@protagolabs.com
huangchengyu24@gmail.com
{soroush, jason.20}@dartmouth.edu

Abstract

We present COVID-Q, a set of 1,690 questions about COVID-19 from 13 sources, which we annotate into 15 question categories and 207 question clusters. The most common questions in our dataset asked about transmission, prevention, and societal effects of COVID, and we found that many questions that appeared in multiple sources were not answered by any FAQ websites of reputable organizations such as the CDC and FDA. We post our dataset publicly at https://github.com/JerryWei03/COVID-Q.

For classifying questions into 15 categories, a BERT baseline scored 58.1% accuracy when trained on 20 examples per category, and for a question clustering task, a BERT + triplet loss baseline achieved 49.5% accuracy. We hope COVID-Q can help either for direct use in developing applied systems or as a domain-specific resource for model evaluation.

1 Introduction

A major challenge during fast-developing pandemics such as COVID-19 is keeping people updated with the latest and most relevant information. Since the beginning of COVID, several websites have created frequently asked questions (FAQ) pages that they regularly update. But even so, users might struggle to find their questions on FAQ pages, and many questions remain unanswered. In this paper, we ask—what are people really asking about COVID, and how can we use NLP to better understand questions and retrieve relevant content?

We present COVID-Q, a dataset of 1,690 questions about COVID from 13 online sources. We annotate COVID-Q by classifying questions into 15 general *question categories*[1] (see Figure 1) and by grouping questions into *question clusters*, for which all questions in a cluster ask the same thing

____

[1] We do not count the "other" category.

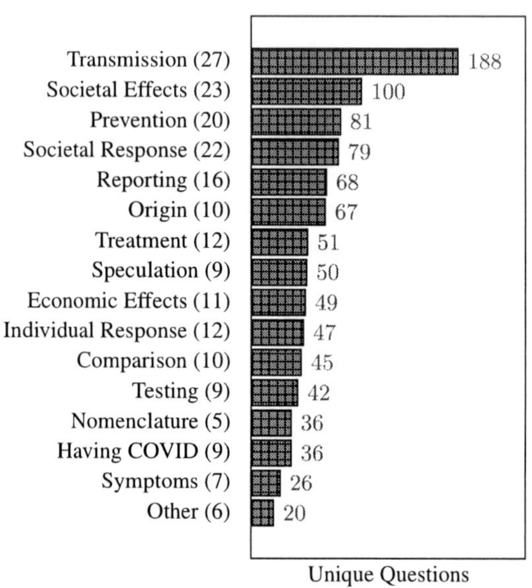

Figure 1: Question categories in COVID-Q, with number of question clusters per category in parentheses.

and can be answered by the same answer, for a total of 207 clusters. Throughout §2, we analyze the distribution of COVID-Q in terms of question category, cluster, and source.

COVID-Q facilitates several question understanding tasks. First, the question categories can be used for a vanilla text classification task to determine the general category of information a question is asking about. Second, the question clusters can be used for retrieval question answering (since the cluster annotations indicate questions of same intent), where given a new question, a system aims to find a question in an existing database that asks the same thing and returns the corresponding answer (Romeo et al., 2016; Sakata et al., 2019). We provide baselines for these two tasks in §3.1 and §3.2. In addition to directly aiding the development of potential applied systems, COVID-Q could also serve as a domain-specific resource for evaluating NLP models trained on COVID data.

Source	Total	Questions		Answers	Questions Removed
		Multi-q-cluster	Single-q-cluster		
Quora	675	501 (74.2%)	174 (25.8%)	0	374
Google Search	173	161 (93.1%)	12 (6.9%)	0	174
github.com/deepset-ai/COVID-QA	124	55 (44.4%)	69 (55.6%)	124	71
Yahoo Search	94	87 (92.6%)	7 (7.4%)	0	34
*Center for Disease Control	92	51 (55.4%)	41 (44.6%)	92	1
Bing Search	68	65 (95.6%)	3 (4.4%)	0	29
*Cable News Network	64	48 (75.0%)	16 (25.0%)	64	1
*Food and Drug Administration	57	33 (57.9%)	24 (42.1%)	57	3
Yahoo Answers	28	13 (46.4%)	15 (53.6%)	0	23
*Illinois Department of Public Health	20	18 (90.0%)	2 (10.0%)	20	0
*United Nations	19	18 (94.7%)	1 (5.3%)	19	6
*Washington DC Area Television Station	16	15 (93.8%)	1 (6.2%)	16	0
*Johns Hopkins University	11	10 (90.9%)	1 (9.1%)	11	1
Author Generated	249	249 (100.0%)	0 (0.0%)	0	0
Total	1,690	1,324 (78.3%)	366 (21.7%)	403	717

Table 1: Distribution of questions in COVID-Q by source. The reported number of questions excludes vague and nonsensical questions that were removed. Multi-q-cluster: number of questions that belonged to a question cluster with at least two questions; Single-q-cluster: number of questions that belonged to a question cluster with only a single question (no other question in the dataset asked the same thing). * denotes FAQ page sources.

2 Dataset Collection and Annotation

Data collection. In May 2020, we scraped questions about COVID from thirteen sources: seven official FAQ websites from recognized organizations such as the Center for Disease Control (CDC) and the Food and Drug Administration (FDA), and six crowd-based sources such as Quora and Yahoo Answers. Table 1 shows the distribution of collected questions from each source. We also post the original scraped websites for each source.

Data cleaning. We performed several pre-processing steps to remove unrelated, low-quality, and nonsensical questions. First, we deleted questions unrelated to COVID and vague questions with too many interpretations (e.g., "Why COVID?"). Second, we removed location-specific and time-specific versions of questions (e.g., "COVID deaths in New York"), since these questions do not contribute linguistic novelty (you could replace "New York" with any state, for example). Questions that only targeted one location or time, however, were not removed—for instance, "Was China responsible for COVID?" was not removed because no questions asked about any other country being responsible for the pandemic. Finally, to minimize occurrences of questions that trivially differ, we removed all punctuation and replaced synonymous ways of saying COVID, such as "coronavirus," and "COVID-19" with "covid." Table 1 also shows the number of removed questions for each source.

Question Cluster [#Questions] (Category)	Example Questions
Pandemic Duration [28] (Speculation)	"Will COVID ever go away?" "Will COVID end soon?" "When COVID will end?"
Demographics: General [26] (Transmission)	"Who is at higher risk?" "Are kids more at risk?" "Who is COVID killing?"
Survivability: Surfaces [24] (Transmission)	"Does COVID live on surfaces?" "Can COVID live on paper?" "Can COVID live on objects?"

Table 2: Most common question clusters in COVID-Q.

Data annotation. We first annotated our dataset by grouping questions that asked the same thing together into question clusters. The first author manually compared each question with existing clusters and questions, using the definition that two questions belong in the same cluster if they have the same answer. In other words, two questions matched to the same question cluster if and only if they could be answered with a common answer. As every new example in our dataset is checked against all existing question clusters, including clusters with only one question, the time complexity for annotating our dataset is $O(n^2)$, where n is the number of questions.

After all questions were grouped into question clusters, the first author gave each question cluster with at least two questions a name summarizing the questions in that cluster, and each question cluster was assigned to one of 15 question categories (as shown in Figure 1), which were conceived during a thorough discussion with the last author. In Table 2, we show the question clusters with the most questions, along with their assigned question categories and some example questions. Figure 2 shows the distribution of question clusters.

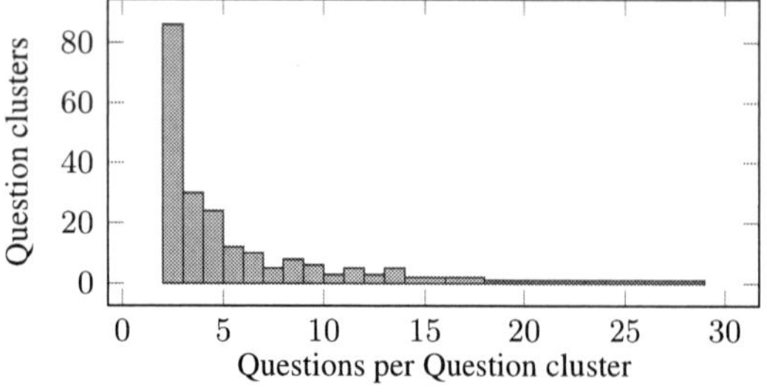

Figure 2: Number of questions per question cluster for clusters with at least two questions. All questions in a question cluster asked roughly the same thing. 120 question clusters had at least 3 questions per cluster, 66 clusters had at least 5 questions per cluster, and 22 clusters had at least 10 questions per cluster.

Annotation quality. We ran the dataset through multiple annotators to improve the quality of our annotations. First, the last author confirmed all clusters in the dataset, highlighting any questions that might need to be relabeled and discussing them with the first author. Of the 1,245 questions belonging to question clusters with at least two questions, 131 questions were highlighted and 67 labels were modified. For a second pass, an external annotator similarly read through the question cluster labels, for which 31 questions were highlighted and 15 labels were modified. Most modifications involved separating a single question cluster that was too broad into several more specific clusters.

For another round of validation, we showed three questions from each of the 89 question clusters with $N_{cluster} \geq 4$ to three Mechanical Turk workers, who were asked to select the correct question cluster from five choices. The majority vote from the three workers agreed with our ground-truth question-cluster labels 93.3% of the time. The three workers unanimously agreed on 58.1% of the questions, for which 99.4% of these unanimous labels agreed with our ground-truth label. Workers were paid $0.07 per question.

Finally, it is possible that some questions could fit in several categories—of 207 clusters, 40 arguably mapped to two or more categories, most frequently the transmission and prevention categories. As this annotation involves some degree of subjectivity, we post formal definitions of each question category with our dataset to make these distinctions more transparent.

Single-question clusters. Interestingly, we observe that for the CDC and FDA frequently asked questions websites, a sizable fraction of questions (44.6% for CDC and 42.1% for FDA) did not ask the same thing as questions from any other source (and therefore formed *single-question clusters*), suggesting that these sources might want adjust the questions on their websites to question clusters that were seen frequently in search engines such as Google or Bing. Moreover, 54.2% of question clusters that had questions from at least two non-official sources went unanswered by an official source. In the Supplementary Materials, Table 7 shows examples of these questions, and conversely, Table 8 shows CDC and FDA questions that did not belong to the same cluster as any other question.

3 Question Understanding Tasks

We provide baselines for two tasks: *question-category classification*, where each question belongs to one of 15 categories, and *question clustering*, where questions asking the same thing belong to the same cluster.

As our dataset is small when split into training and test sets, we manually generate an additional *author-generated* evaluation set of 249 questions. For these questions, the first author wrote new questions for question clusters with 4 or 5 questions per cluster until those clusters had 6 questions. These questions were checked in the same fashion as the real questions. For clarity, we only refer to them in §3.1 unless explicitly stated.

3.1 Question-Category Classification

The *question-category classification* task assigns each question to one of 15 categories shown in Figure 1. For the train-test split, we randomly choose 20 questions per category for training (as the smallest category has 26 questions), with the remaining questions going into the test set (see Table 3).

Question Categories	15
Training Questions per Category	20
Training Questions	300
Test Questions (Real)	668
Test Questions (Generated)	238

Table 3: Data split for *question-category classification*.

We run simple BERT (Devlin et al., 2019) feature-extraction baselines with question representations obtained by average-pooling. For this task, we use two models: (1) SVM and (2) cosine-similarity based k-nearest neighbor classification (k-NN) with $k = 1$. As shown in Table 4, the SVM marginally outperforms k-NN on both the real and generated evaluation sets. Since our dataset is small, we also include results from using data augmentation (Wei and Zou, 2019). Figure 4 (Supplementary Materials) shows the confusion matrix for BERT-feat: SVM + augmentation for this task.

Model	Real Q	Generated Q
BERT-feat: k-NN	47.8	52.1
+ augmentation	47.3	52.5
BERT-feat: SVM	52.2	53.4
+ augmentation	58.1	58.8

Table 4: Performance of BERT baselines (accuracy in %) on *question-category classification* with 15 categories and 20 training examples per category.

3.2 Question Clustering

Of a more granular nature, the *question clustering* task asks, given a database of known questions, whether a new question asks the same thing as an existing question in the database or whether it is a novel question. To simulate a potential applied setting as much as possible, we use all questions clusters in our dataset, including clusters containing only a single question. As shown in Table 5, we make a 70%–30% train–test split by class.[2]

Training Questions	920
Training Clusters	460
Test Questions	437
Test Clusters	320
Test Questions from multi-q-clusters	323
Test Questions from single-q-clusters	114

Table 5: Data split for *question clustering*.

In addition to the k-NN baseline from §3.1, we also evaluate a simple model that uses a triplet loss function to train a two-layer neural net on BERT features, a method introduced for facial recognition (Schroff et al., 2015) and now used in NLP for few-shot learning (Yu et al., 2018) and answer selection (Kumar et al., 2019). For evaluation, we compute a single accuracy metric that requires a question to be either correctly matched to a cluster in the database or to be correctly identified as a novel question. Our baseline models use thresholding to determine

---

[2]For clusters with two questions, one question went into the training set and one into the test set. 70% of single-question clusters went into the training set and 30% into the test set.

Model	Accuracy (%)	
	Top-1	Top-5
BERT-feat: k-NN	39.6	58.8
+ augmentation	39.6	59.0
BERT-feat: triplet loss	47.7	66.9
+ augmentation	49.5	69.4

Table 6: Performance of BERT baselines on *question clustering* involving 207 clusters.

whether questions were in the database or novel. Table 6 shows the accuracy from the best threshold for both these models, and Supplementary Figure 3 shows their accuracies for different thresholds.

4 Discussion

Use cases. We imagine several use cases for COVID-Q. Our question clusters could help train and evaluate retrieval-QA systems, such as `covid.deepset.ai` or `covid19.dialogue.co`, which, given a new question, aim to retrieve the corresponding QA pair in an existing database. Another relevant context is query understanding, as clusters identify queries of the same intent, and categories identify queries asking about the same topic. Finally, COVID-Q could be used broadly to evaluate COVID-specific models—our baseline (Huggingface's `bert-base-uncased`) does not even have *COVID* in the vocabulary, and so we suspect that models pre-trained on scientific or COVID-specific data will outperform our baseline. More related areas include COVID-related query expansion, suggestion, and rewriting.

Limitations. Our dataset was collected in May 2020, and we see it as a snapshot in time of questions asked up until then. As the COVID situation further develops, a host of new questions will arise, and the content of these new questions will potentially not be covered by any existing clusters in our dataset. The question categories, on the other hand, are more likely to remain static (i.e., new questions would likely map to an existing category), but the current way that we came up with the categories might be considered subjective—we leave that determination to the reader (refer to Table 9 or the raw dataset on Github). Finally, although the distribution of questions per cluster is highly skewed (Figure 2), we still provide them at least as a reference for applied scenarios where it would be useful to know the number of queries asking the same thing (and perhaps how many answers are needed to answer the majority of questions asked).

References

Muhammad Abdul-Mageed, AbdelRahim Elmadany, Dinesh Pabbi, Kunal Verma, and Rannie Lin. 2020. Mega-cov: A billion-scale dataset of 65 languages for covid-19. *ArXiv*, abs/2005.06012. https://arxiv.org/pdf/2005.06012.pdf.

Emily Chen, Kristina Lerman, and Emilio Ferrara. 2020. Covid-19: The first public coronavirus twitter dataset. *ArXiv*, abs/2003.07372. https://arxiv.org/pdf/2003.07372.pdf.

Jacob Devlin, Ming-Wei Chang, Kenton Lee, and Kristina Toutanova. 2019. BERT: Pre-training of deep bidirectional transformers for language understanding. In *Proceedings of the 2019 Conference of the North American Chapter of the Association for Computational Linguistics: Human Language Technologies, Volume 1 (Long and Short Papers)*, pages 4171–4186, Minneapolis, Minnesota. Association for Computational Linguistics. https://www.aclweb.org/anthology/N19-1423.pdf.

Zhiwei Gao, Shuntaro Yada, Shoko Wakamiya, and Eiji Aramaki. 2020. Naist covid: Multilingual covid-19 twitter and weibo dataset. *ArXiv*, abs/2004.08145. https://arxiv.org/pdf/2004.08145.pdf.

Ting-Hao Huang, Chieh-Yang Huang, Chien-Kuang Cornelia Ding, Yen-Chia Hsu, and C. Lee Giles. 2020. Coda-19: Reliably annotating research aspects on 10, 000+ cord-19 abstracts using non-expert crowd. *ArXiv*, abs/2005.02367. https://arxiv.org/pdf/2005.02367.pdf.

Bennett Kleinberg, Isabelle van der Vegt, and Maximilian Mozes. 2020. Measuring emotions in the covid-19 real world worry dataset. *ArXiv*, abs/2004.04225. https://arxiv.org/pdf/2004.04225.pdf.

Sawan Kumar, Shweta Garg, Kartik Mehta, and Nikhil Rasiwasia. 2019. Improving answer selection and answer triggering using hard negatives. In *Proceedings of the 2019 Conference on Empirical Methods in Natural Language Processing and the 9th International Joint Conference on Natural Language Processing (EMNLP-IJCNLP)*, pages 5911–5917. Association for Computational Linguistics. "https://www.aclweb.org/anthology/D19-1604".

Salvatore Romeo, Giovanni Da San Martino, Alberto Barrón-Cedeño, Alessandro Moschitti, Yonatan Belinkov, Wei-Ning Hsu, Yu Zhang, Mitra Mohtarami, and James Glass. 2016. Neural attention for learning to rank questions in community question answering. In *Proceedings of COLING 2016, the 26th International Conference on Computational Linguistics: Technical Papers*, pages 1734–1745, Osaka, Japan. The COLING 2016 Organizing Committee. https://www.aclweb.org/anthology/C16-1163.pdf.

Wataru Sakata, Tomohide Shibata, Ribeka Tanaka, and Sadao Kurohashi. 2019. FAQ retrieval using query-question similarity and bert-based query-answer relevance. *CoRR*, abs/1905.02851. https://arxiv.org/pdf/1905.02851.pdf.

A. Sarker, S. Lakamana, William E. Hogg, Allen Xie, Mohammed Ali Al-garadi, and Yc Yang. 2020. Self-reported covid-19 symptoms on twitter: An analysis and a research resource. In *medRxiv*. https://www.medrxiv.org/content/10.1101/2020.04.16.20067421v3.full.pdf.

Florian Schroff, Dmitry Kalenichenko, and James Philbin. 2015. Facenet: A unified embedding for face recognition and clustering. *CoRR*, abs/1503.03832. http://arxiv.org/abs/1503.03832.

Jason Wei and Kai Zou. 2019. EDA: easy data augmentation techniques for boosting performance on text classification tasks. *Proceedings of the 2019 Conference on Empirical Methods in Natural Language Processing and the 9th International Joint Conference on Natural Language Processing (EMNLP-IJCNLP)*. http://dx.doi.org/10.18653/v1/D19-1670.

Mo Yu, Xiaoxiao Guo, Jinfeng Yi, Shiyu Chang, Saloni Potdar, Yu Cheng, Gerald Tesauro, Haoyu Wang, and Bowen Zhou. 2018. Diverse few-shot text classification with multiple metrics. In *Proceedings of the 2018 Conference of the North American Chapter of the Association for Computational Linguistics: Human Language Technologies, Volume 1 (Long Papers)*, pages 1206–1215. Association for Computational Linguistics. https://www.aclweb.org/anthology/N18-1109.

Koosha Zarei, Reza Farahbakhsh, Noel Crespi, and Gareth Tyson. 2020. A first instagram dataset on covid-19. *ArXiv*, abs/2004.12226. https://arxiv.org/pdf/2004.12226.pdf.

5 Supplementary Materials

5.1 Question Clustering Thresholds

For the question clustering task, our models used simple thresholding to determine whether a question matched an existing cluster in the database or was novel. That is, if the similarity between a question and its most similar question in the database was lower than some threshold, then the model predicted that it was a novel question. Figure 3 shows the accuracy of the k-NN and triplet loss models at different thresholds.

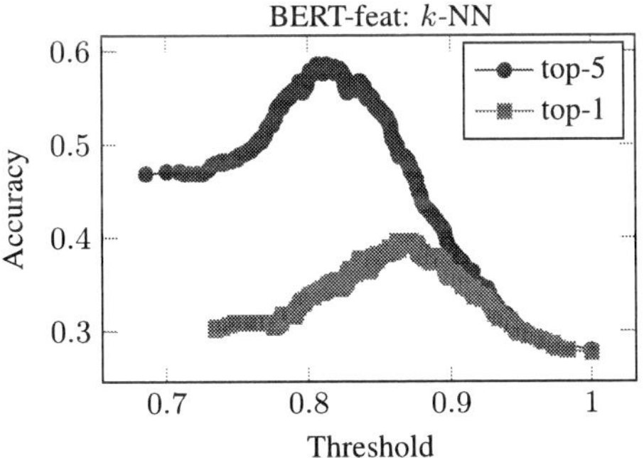

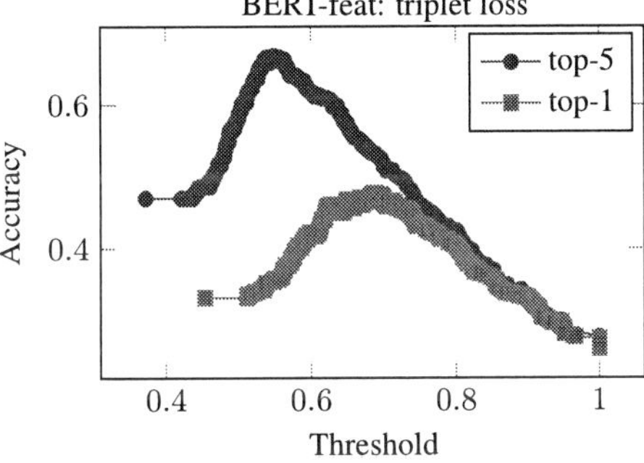

Figure 3: Question clustering accuracy for k-NN and triplet loss models at different thresholds. If a given test question had a similarity that was less than the threshold, then it was classified as a novel question (i.e., not in the database of known questions). When the threshold was too high, performance dropped because too many questions were classified as novel. When the threshold was too low, performance dropped because the model attempted to match too many test questions to existing clusters in the database.

5.2 Question-Category Classification Error Analysis

Figure 4 shows the confusion matrix for our SVM classifier on the question-category classification task on the test set of real questions. Categories that were challenging to distinguish were *Transmission* and *Having COVID* (34% error rate), and *Having COVID* and *Symptoms* (33% error rate).

5.3 Further Dataset Details

Question mismatches. Table 7 shows example questions from at least two non-official sources that went unanswered by an official source. Table 8 shows example questions from the FDA and CDC FAQ websites that did not ask the same thing as any other questions in our dataset.

Question Cluster	$N_{cluster}$	Example Questions
Number of Cases	21	"Are COVID cases dropping?" "Have COVID cases peaked?" "Are COVID cases decreasing?"
Mutation	19	"Has COVID mutated?" "Did COVID mutate?" "Will COVID mutate?"
Lab Theory	18	"Was COVID made in a lab?" "Was COVID manufactured?" "Did COVID start in a lab?"

Table 7: Questions appearing in multiple sources that were unanswered by official FAQ websites.

Example questions. Table 9 shows example questions from each of the 15 question categories.

Corresponding answers. The FAQ websites from reputable sources (denoted with * in Table 1) provide answers to their questions, and so we also provide them as an auxiliary resource. Using these answers, 23.8% of question clusters have at least one corresponding answer. We caution against using these answers in applied settings, however, because information on COVID changes rapidly.

Additional data collection details. In terms of how questions about COVID were determined, for FAQ websites from official organizations, we considered all questions, and for Google, Bing, Yahoo, and Quora, we searched the keywords "COVID" and "coronavirus."

As for synonymous ways of saying COVID, we considered "SARS-COV-2," "coronavirus," "2019-nCOV," "COVID-19," and "COVID19."

Other COVID-19 datasets. We encourage researchers to also explore other COVID-19 datasets: tweets streamed since January 22 (Chen et al., 2020), location-tagged tweets in 65 languages (Abdul-Mageed et al., 2020), tweets of COVID symptoms (Sarker et al., 2020), a multi-lingual Twitter and Weibo dataset (Gao et al., 2020), an Instagram dataset (Zarei et al., 2020), emotional responses to COVID (Kleinberg et al., 2020), and annotated research abstracts (Huang et al., 2020).

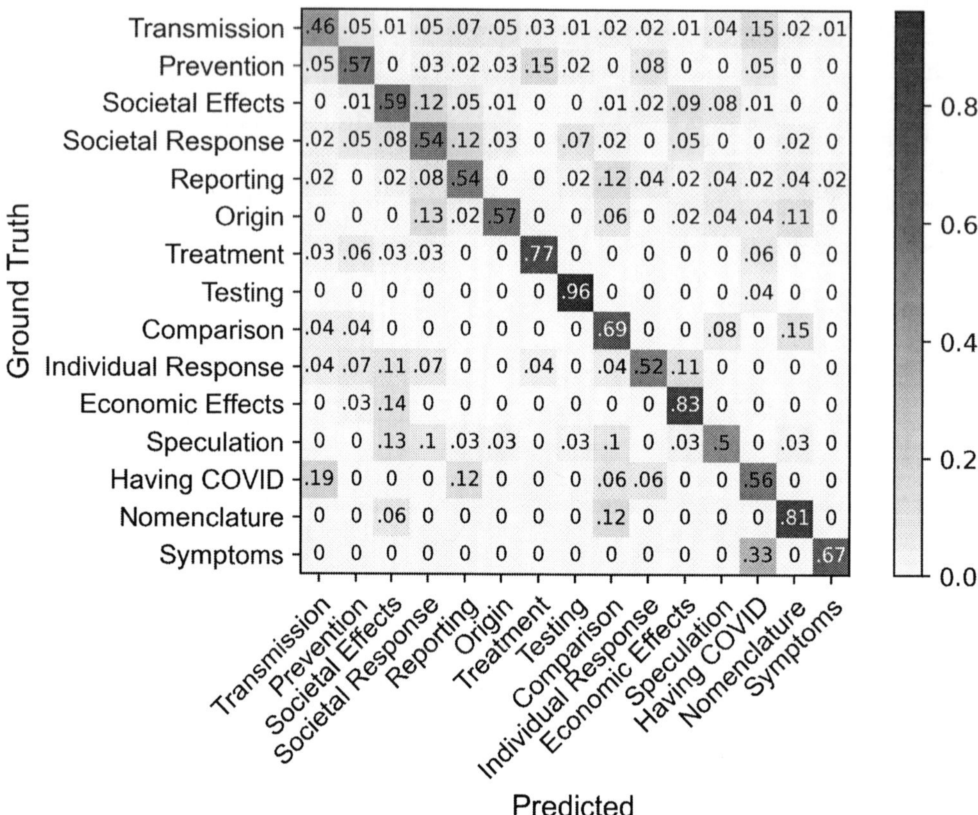

Figure 4: Confusion matrix for BERT-feat: SVM predictions on the question-category classification task.

Question	Food and Drug Administration Closest Matches from BERT
"Can I donate convalescent plasma?"	"Why is convalescent plasma being investigated to treat COVID?" "Can I make my own hand sanitizer?" "What are suggestions for things to do in the COVID quarantine?"
"Where can I report websites selling fraudulent medical products?"	"What kind of masks are recommended to protect healthcare workers from COVID exposure?" "Where can I get tested for COVID?" "How do testing kits for COVID detect the virus?"

Question	Center for Disease Control Closest Matches from BERT
"What is the difference between cleaning and disinfecting?"	"How effective are alternative disinfection methods?" "Why has Trump stated that injecting disinfectant will kill COVID in a minute?" "Should I spray myself or my kids with disinfectant?"
"How frequently should facilities be cleaned to reduce the potential spread of COVID?"	"What is the survival rate of those infected by COVID who are put on a ventilator?" "What kind of masks are recommended to protect healthcare workers from COVID exposure?" "Will warm weather stop the outbreak of COVID?"

Table 8: Questions from the Food and Drug Administration (FDA) and Center for Disease Control (CDC) FAQ websites that did not ask the same thing as any questions from other sources.

Category	Example Questions
Transmission	"Can COVID spread through food?" "Can COVID spread through water?" "Is COVID airborne?"
Societal Effects	"In what way have people been affected by COVID?" "How will COVID change the world?" "Do you think there will be more racism during COVID?"
Prevention	"Should I wear a facemask?" "How can I prevent COVID?" "What disinfectants kill the COVID virus?"
Societal Response	"Have COVID checks been issued?" "What are the steps that a hospital should take after COVID outbreak?" "Are we blowing COVID out of proportion?"
Reporting	"Is COVID worse than we are being told?" "What is the COVID fatality rate?" "What is the most reliable COVID model right now?"
Origin	"Where did COVID originate?" "Did COVID start in a lab?" "Was COVID a bioweapon?"
Treatment	"What treatments are available for COVID?" "Should COVID patients be ventilated?" "Should I spray myself or my kids with disinfectant?"
Speculation	"Was COVID predicted?" "Will COVID return next year?" "How long will we be on lockdown for COVID?"
Economic Effects	"What is the impact of COVID on the global economy?" "What industries will never be the same because of COVID?" "Why are stock markets dipping in response to COVID?"
Individual Response	"How do I stay positive with COVID?" "What are suggestions for things to do in the COVID quarantine?" "Can I still travel?"
Comparison	"How are COVID and SARS-COV similar?" "How can I tell if I have the flu or COVID?" "How does COVID compare to other viruses?"
Testing	"How COVID test is done?" "Are COVID tests accurate?" "Should I be tested for COVID?"
Nomenclature	"Should COVID be capitalized?" "What COVID stands for?" "What is the genus of the SARS-COVID?"
Having COVID	"How long does it take to recover?" "How COVID attacks the body?" "How long is the incubation period for COVID?"
Symptoms	"What are the symptoms of COVID?" "Which COVID symptoms come first?" "Do COVID symptoms come on quickly?"

Table 9: Sample questions from each of the 15 question categories.

Jennifer for COVID-19: An NLP-Powered Chatbot Built for the People and by the People to Combat Misinformation

Yunyao Li[1,*]   Tyrone Grandison[2,*]   Patricia Silveyra[3,*]   Ali Douraghy[4]
Xinyu Guan[5]   Thomas Kieselbach[6]   Chengkai Li[7]   Haiqi Zhang[7]

[1] IBM Research - Almaden [2] The Data-Driven Institute [3] University of North Carolina - Chapel Hill
[4] The National Academies of Sciences, Engineering and Medicine [5] Yale University
[6] Umeå University [7] University of Texas - Arlington

{yunyaoli@us.ibm.com, tgrandison@data-driven.institute, patry@email.unc.edu, adouraghy@nas.edu}

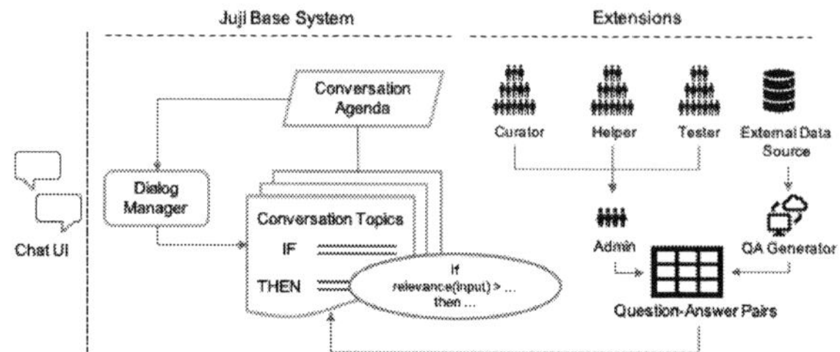

Figure 1: Architecture Overview of Jennifer

Just as SARS-CoV-2 continues to infect a growing number of people around the world, harmful misinformation about the outbreak also continues to spread. We designed and built `Jennifer` chatbot to provide easily accessible information from reliable resources to answer questions related to the current COVID-19 pandemic. It covers a wide variety of topics, from case statistics to best practices for disease prevention and management.

With `Jennifer`, we hope to learn whether public information from reputable sources could be more effectively organized and shared in the wake of a crisis as well as to understand issues that the public are most immediately curious about (New Voices, 2020). Our core design considerations are:

- **Rapid Development**: `Jennifer` should be built within a short amount of time to win the race against fast-spreading misinformation.
- **Ease of Access**: `Jennifer` should provide information to the general public in an easily accessible manner across different platforms.
- **Ease of Maintenance**: `Jennifer` should be maintainable by a diverse group of volunteers.
- **Quality Assurance**: `Jennifer` should provide information from reputable sources in a consumable and empathetic manner, and maintain a rigorous process to ensure its quality adn accuracy.
- **Extensibility**: `Jennifer` should be easily extensible to expand its capability with minimal effort.

`Jennifer` depends on the Juji (Juji, 2020) base system for dialog management (Figure 1). Given a question, Juji uses a pre-trained machine learning model to identify relevant questions with known answers and returns an answer or a follow-up question. The main capabilities of `Jennifer` come from the Question-Answer(QA) pairs that are either manually curated by our volunteers or auto-generated via manually-curated templates.

The first version of `Jennifer` was designed and released within 24 hours on March 8, 2020. Since then, over 160 volunteers from 141 institutions around the globe recruited through the New Voices' network have helped make updates to the chatbot to ensure that its content reflects the latest available information from trusted sources. `Jennifer` is available on the Web (`http://bit.ly/jenniferai`), Facebook[1] and embedded in two fact-checking systems.[2] As of June 18, 2020, `Jennifer` has been asked 1,480 questions (excluding questions selected via menus) and answered 1,059 of them (a response rate of 71%), with an average engagement duration of 3 minutes and 15 seconds. We have released COQB-19 (COVID-19 Question Bank)[3], including 3,924 COVID-19-related questions in 944 groups, gathered from our users and volunteers.

References

Juji. 2020. Juji document for chatbot designers. `https://docs.juji.io/`. [Online; accessed 14-June-2020].

New Voices. 2020. How the us must respond to the covid-19 pandemic. `https://blogs.scientificamerican.com/observations/how-the-us-must-respond-to-the-covid-19-pandemic/`. [Online; accessed June-2020].

---

*denotes equal contribution

[1] Facebook `http://fb.me/JenniferCOVIDAI`
[2] `https://coronacheck.eurecom.fr/en` and `https://idir.uta.edu/covid-19/`
[3] `https://www.newvoicesnasem.org/data-downloads`

A Natural Language Processing System for National COVID-19 Surveillance in the US Department of Veterans Affairs

Alec B Chapman[1,2]**, Kelly S Peterson**[1,2]**, Augie Turano**[3]**, Tamára L Box**[4]**,
Katherine S Wallace**[5]**, Makoto Jones**[1,2]

[1] Veterans Affairs (VA) Salt Lake City Health Care System
[2] Division of Epidemiology, University of Utah
[3] VA Office of EHR Modernization
[4] VA Office of Clinical Systems Development and Evaluation (CSDE)
[5] VA Office of Biosurveillance, VA Central Office, Washington, DC

Abstract

Timely and accurate accounting of positive cases has been an important part of the response to the COVID-19 pandemic. While most positive cases within Veterans Affairs (VA) are identified through structured laboratory results, some patients are tested or diagnosed outside VA so their clinical status is documented only in free-text narratives. We developed a Natural Language Processing pipeline for identifying positively diagnosed COVID-19 patients and deployed this system to accelerate chart review. As part of the VA national response to COVID-19, this process identified 6,360 positive cases which did not have corresponding laboratory data. These cases accounted for 36.1% of total confirmed positive cases in VA to date. With available data, performance of the system is estimated as 82.4% precision and 94.2% recall. A public-facing implementation is released as open source and available to the community.

1 Introduction

A robust pandemic response is contingent on timely and accurate information (Morse 2012). During the COVID-19 pandemic, public health institutions have established surveillance systems to monitor and track case counts over time.

COVID-19 is typically diagnosed using laboratory tests. The test results are frequently used as a source for surveillance systems. However, such systems typically only capture laboratory results from the same healthcare system. Patients may also be diagnosed with COVID-19 in the community, such as in external hospital networks or drive-through testing. These patients may potentially be missed by laboratory-based surveillance methods, leading to these patients not being represented in overall case counts.

Patient health information needed for biosurveillance is often recorded in free-text narratives in the Electronic Health Record (EHR) (Chapman et al. 2011), offering an alternative source of COVID-19 status when structured lab evidence is absent.

In this work we developed a Natural Language Processing (NLP) system to extract potential positive COVID-19 cases from clinical text within the Department of Veterans Affairs (VA). Following review by a clinical expert, positively identified patients are included in official VA surveillance counts. Since the VA EHR includes data from hospitals and clinics across the United States, this system enables a unique capability for collecting data for national surveillance purposes.

2 Background

Manual information gathering draws effort away from patient care priorities and can impede timely and effective responses to public health threats. Automated approaches for processing clinical notes have been applied for public health purposes when data is needed as quickly as possible.

Gesteland et al (2003) developed an automated syndromic surveillance system using clinical text to identify anomalies in symptoms as rapidly as possible. Several examples in the literature have utilized clinical text including chief complaints to perform early detection of infectious disease (Brillman et al. 2005; Chapman, Dowling, and Wagner 2004; Ivanov et al. 2003; Matheny et al. 2012; Pineda et al. 2015).

Typical data sources for COVID-19 surveillance include government announcements, scientific publications, and news articles (Xu et al. 2020) Most literature to date for NLP related to COVID-19 has involved public data sources such as research publications (Wang et al. 2020). Others have examined social media sources including Twitter to examine sentiment or misinformation related to the virus (Rajput, Grover, and Rathi 2020; Singh et al. 2020). In this work, the objective was to identify the diagnosis of COVID-19 in clinical documents to report complete case counts of the disease for public health surveillance in VA.

3 Methods

3.1 Dataset

Veterans Health Administration (VHA) includes medical centers and clinics across the United States [1]. The VA Corporate Data Warehouse (CDW) includes electronic clinical data for these sites in a unified architecture. This work included clinical data in 2020 between January 1 and June 15.

3.2 NLP Pipeline

The primary objective of our NLP system is to classify whether a clinical document contains a positive COVID-19 case. To do this, we designed a rule-based pipeline which extracted target entities related to COVID-19, asserted certain attributes for each entity, and finally classified documents as either positive or negative based on the entities within the document. We prioritized minimizing false negatives in order to identify as many positive cases as possible. However, as the volume of data increased, it became important to reduce false positives in order to minimize manual chart review.

The pipeline was implemented in Python using the spaCy framework[2]. All processing steps except for tokenization, part-of-speech tagging, and dependency parsing were implemented using custom spaCy components, a feature available in version 2.0 and later. Each component may contain its own rules or knowledge base. Several components are available as part of medSpaCy[3], an open source project for clinical NLP using spaCy, and a publicly available version of the pipeline is released on GitHub[4].

The following describes each of the custom components in the pipeline, shown visually in Appendix A:

- **Preprocessor**: Modifies the underlying text before text processing. This step removes semi-structured templated texts and questionnaires which can cause false positives and replaces certain abbreviations and misspellings to simplify later processing steps.
- **Target Matcher:** Extracts entities related to COVID-19 based on linguistic patterns. This includes terms such as *"COVID-19"*, *"novel coronavirus"*, *"ncov"*, and *"SARS-COV-2"*.
- **Context:** Identifies semantic modifiers and attributes such as negation, uncertainty, and experiencer. This step was performed using cycontext [5], a spaCy implementation of the ConText algorithm (Chapman, Dowling, and Chu 2007). Figure 1 shows a visualization of the ConText algorithm.
- **Sectionizer:** Detects section boundaries in the text, such as *"Visit Diagnoses"* or *"Past Medical History"*.
- **Postprocessor:** Modifies or removes entities based on business logic. This component allows the pipeline to handle edge cases or more complex logic using the results of previous components.
- **Document Classifier:** Assigns a label of "Positive" or "Negative" to each document based on the entities and attributes extracted from the text.

The following is a brief description of classification logic at both entity level and document level. Entities are excluded if any of the following attributes are present:

- Uncertain
- Negated
- Experienced by someone other than the patient

Entities are marked as "positive" when any of the following conditions are met:

- Associated with a positive modifier, such as *"diagnosed with"* or *"is positive"*
- Occurring in certain sections of a note, such as *"Diagnoses:"*
- Mentioned with a specific associated condition, such as *"COVID-19 pneumonia"*

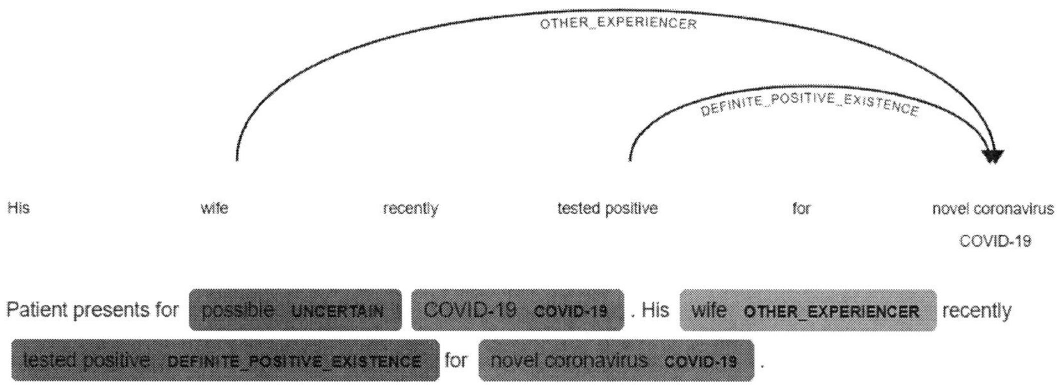

Figure 1. Visualizations provided in medSpaCy allowed us to view the output of our system and inspect linguistic patterns in the text. Target and modifier concepts are highlighted in text and arrows between them show relationships indicating whether the patient experienced COVID-19.

Based on the entities and corresponding attributes, we then classify the document as "Positive" or "Negative". In our current implementation, a document is classified as "Positive" if it has at least one positive, non-excluded entity.

3.3 Deployment

Our system was deployed to process clinical notes in VA CDW beginning January 21, 2020, the day after the first case was confirmed in the United States (Holshue et al. 2020). All documents containing keywords related to COVID-19 were included in document processing. Documents were retrieved and processed regularly to facilitate daily operations.

3.4 Clinical Review

When a patient's document was classified by text processing as positive, the document was reviewed by a clinical validator. Using an internally developed web-based tool, reviewers viewed a marked-up summary of the processed clinical documents. If the patient fit a clinical definition of COVID-19, the reviewer accepted the suggestion and the patient was added to VA's COVID-19 counts.

Due to an increasing volume of data and limited resources for review, later iterations accelerated validation and improved precision by assigning documents to "High" and "Low" priority groups using other indicators such as a relevant ICD-10 code. This allowed reviewers to prioritize review of those patients who were likely to be valid cases and to minimize the review of false positives.

4   Results

4.1 Document Processing

Keywords such as *coronavirus, novel coronavirus, COVID-19, SARS-CoV-2,* and others were found in 17 million documents in VA CDW between

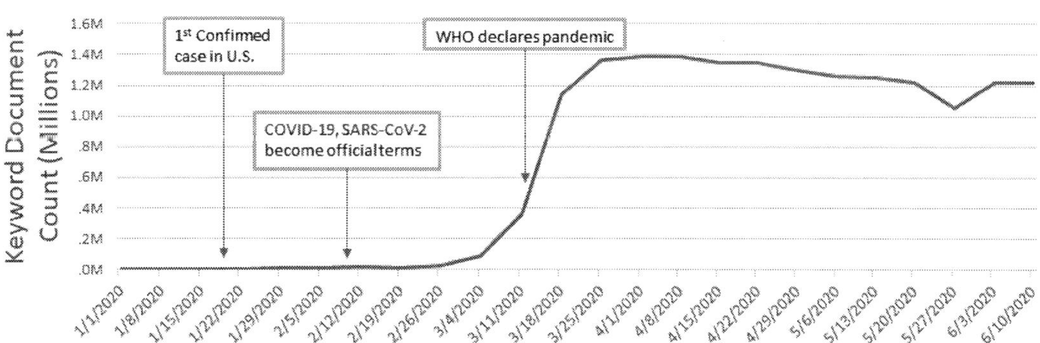

Figure 2. Frequency of documents matching COVID-19 related keywords from January through June 15, 2020. Some key dates are marked for reference.

January 1 and June 15, 2020. The median document length of this document set was 1,383 characters. Figure 2 shows the weekly volume of documents matching these keywords.

The phrase *novel coronavirus* was first observed in clinical notes the week of January 15. On February 11, 2020, World Health Organization (WHO) announced terminology of *SARS-CoV-2* for the virus and *COVID-19* as the disease it causes (World Health Organization 2020a). On March 11, WHO declared the COVID-19 situation as a pandemic (World Health Organization 2020b). In our dataset, the term COVID-19 occurred nearly 50,000 times the week of March 11 and increased to over 250,000 mentions the following week.

As of June 15, 2020, our system had processed documents from 3.6 million patients. Table 1 presents several illustrations of example text processed and classified by our system. After clinical review, a total of 6,360 patients without laboratory evidence were confirmed to be positive for COVID-19. This accounted for 36.1% of the total 17,624 positive cases identified in VA at the time.

Text Classifications	
Positive	"Patient admitted to hospital for respiratory failure secondary to ***COVID-19***." "Diagnoses: ***COVID-19 B34.9***" "The patient reports that they have been diagnosed with ***COVID-19***."
Negative	"Requested that patient be screened for ***COVID-19*** via telephone." "Studies have shown that some ***COVID-19*** patients have prolonged baseline." "Has the patient been diagnosed with ***COVID-19***? Y/N"

Table 1. Examples of positive and negative classified text.

4.2 System Performance

To evaluate the performance of our pipeline, we estimated precision and recall. Due to constraints, we calculated precision at a document level and recall at a patient level.

For precision, we manually reviewed 500 randomly selected documents classified as positive with an entry date on or later than May 1. We considered a document a true positive if the patient was stated to have been positive for COVID-19 and thus appropriate to review for validation.

Measuring recall is more complicated as the actual number of positive cases is not known. To estimate recall, we evaluated performance of our system for patients with positive laboratory results and at least one document containing previously mentioned keywords. We considered recall to be the percentage of these patients who had at least one document classified as positive by our system. All positive COVID-19 laboratory results completed between May 1 and June 15 were included in this analysis.

Our review yielded an estimated document-level precision of 82.4%. Estimated patient-level recall was 94.2%. Appendix B shows examples and explanations of incorrectly classified texts. One common cause of false positives was template texts such as screenings or educational information which contained phrases such as "confirmed COVID-19" but did not actually signify that the patient was positive. Several errors were referring to COVID-19 practices or the pandemic more generally, such as "COVID-19 infection control protocols". Other errors were caused by incorrectly linked targets and modifiers, resulting in marking a non-positive entity as positive or failing to mark an entity as excluded.

One source of false negatives was positive modifiers which were not linked to mentions of COVID-19. The scope for linking targets and modifiers was set to be one sentence based upon observation that linguistic modifiers typically occurred in the same sentence as a target concept. This error can be propagated by text formatting such as erroneous new lines which cause incorrect sentence splitting.

5   Discussion

In this work we described the development and application of a Natural Language Processing system for COVID-19 surveillance in a national healthcare system in the United States. We demonstrated that NLP combined with clinical review can be leveraged to improve surveillance for COVID-19. Within the VA surveillance system, over one third of total known cases were identified by NLP and clinical review, with the remainder being identified through structured laboratory data. This capability validated that NLP can provide significant value to such a surveillance system, which requires a timely and sensitive case count.

Our system achieved high recall while still maintaining acceptable precision. Leveraging a rule-based system allowed defining narrow and specific criteria for what is extracted. Rules were iteratively developed to filter out irrelevant documents while still identifying positive cases.

Additionally, the flexibility of a rule-based system allowed us to add new examples and adapt to new concepts as they emerged. This was critical in the COVID-19 response, as the pandemic remains a dynamic and evolving situation. For example, the terms *COVID-19* and *SARS-CoV-2* were not announced until weeks after the surveillance system had been deployed, but requirements dictated immediate addition to our system. Similarly, changes in the clinical documentation such as new clinical concerns and semi-structured template texts required quick response and modification.

Due to the continuously changing nature of COVID-19, we required a system which permitted rapid and flexible development. While other mature clinical NLP systems exist, such as cTAKES and CLAMP (Savova et al. 2010; Soysal et al. 2018), we elected to develop this system using the features and flexibility of the spaCy framework. Rapid iteration permitted reviewing documents for errors, directly making changes to rules, and then evaluating them without compiling or reloading. Visualizations such as Figure 1 were useful to troubleshoot rule development and understand the linguistic patterns.

One limitation of this work is the evaluation of system performance. Our primary objective in this effort was to serve Veterans and provide complete public health reporting. The goal of chart review was to identify all positive patients rather than to create a reference set. Precision and recall metrics presented here are estimates using sampling and available structured data.

In future work, we plan to evaluate machine learning methods to improve identification of positive cases. A machine learning classifier could potentially improve our current system by improving document classification accuracy and identifying high-probability cases for review. This

was not feasible in early stages of the response since there were very few known cases and no existing reference set. We have now identified thousands of possible cases which could be included in a training set for a supervised classifier. However, as stated previously, our clinical review did not equate to creating a reference set. Specifically, clinical reviewers did not always assign negative labels to reviewed cases which would be needed for training a supervised model. However, we believe that with additional validation and review, a machine learning classifier has the potential to augment our system's performance.

6   Conclusion

We have developed a text processing pipeline and utilized it to perform accelerated review of COVID-19 status in clinical documents. This approach was dynamic and allowed us to adapt to an evolving situation where vocabulary and clinical understanding continued to emerge with high data volume. Rapid implementation and iteration permitted reaction to shifting clinical documentation and evidence. This pipeline accelerated review of patient charts such that 36.1% of confirmed positive cases in a VA surveillance system were identified using this capability.

Acknowledgments

We thank Christopher Mannozzi, Gary Roselle, Joel Roos, Joseph Francis, Julia Lewis, Richard Pham, Shantini Gamage, VA Business Intelligences Services Line (BISL), VA Office of Clinical Systems Development and Evaluation (CSDE), VHA Healthcare Operations Center (HOC), VHA Office of Analytics and Performance Integration (API), and VA Informatics and Computing Infrastructure (VINCI) Applied NLP.

We also thank the members of CSDE BASIC (Biosurveillance, Antimicrobial Stewardship, and Infection Control) for their invaluable contributions to this work.

References

Brillman, Judith C., Tom Burr, David Forslund, Edward Joyce, Rick Picard, and Edith Umland. 2005. "Modeling Emergency Department Visit Patterns for Infectious Disease Complaints: Results and

Application to Disease Surveillance." *BMC Medical Informatics and Decision Making* 5(1):4.

Chapman, Wendy, John Dowling, and David Chu. 2007. "ConText: An Algorithm for Identifying Contextual Features from Clinical Text." Pp. 81–88 in *Biological, translational, and clinical language processing*.

Chapman, Wendy W., John N. Dowling, and Michael M. Wagner. 2004. "Fever Detection from Free-Text Clinical Records for Biosurveillance." *Journal of Biomedical Informatics* 37(2):120–27.

Chapman, Wendy W., Adi V Gundlapalli, Brett R. South, and John N. Dowling. 2011. "Natural Language Processing for Biosurveillance." Pp. 279–310 in *Infectious Disease Informatics and Biosurveillance*. Springer.

Gesteland, Per H., Reed M. Gardner, Fu-Chiang Tsui, Jeremy U. Espino, Robert T. Rolfs, Brent C. James, Wendy W. Chapman, Andrew W. Moore, and Michael M. Wagner. 2003. "Automated Syndromic Surveillance for the 2002 Winter Olympics." *Journal of the American Medical Informatics Association* 10(6):547–54.

Holshue, Michelle L., Chas DeBolt, Scott Lindquist, Kathy H. Lofy, John Wiesman, Hollianne Bruce, Christopher Spitters, Keith Ericson, Sara Wilkerson, and Ahmet Tural. 2020. "First Case of 2019 Novel Coronavirus in the United States." *New England Journal of Medicine*.

Ivanov, Oleg, Per H. Gesteland, William Hogan, Michael B. Mundorff, and Michael M. Wagner. 2003. "Detection of Pediatric Respiratory and Gastrointestinal Outbreaks from Free-Text Chief Complaints." P. 318 in *AMIA Annual Symposium Proceedings*. Vol. 2003. American Medical Informatics Association.

Matheny, Michael E., Fern FitzHenry, Theodore Speroff, Jennifer K. Green, Michelle L. Griffith, Eduard E. Vasilevskis, Elliot M. Fielstein, Peter L. Elkin, and Steven H. Brown. 2012. "Detection of Infectious Symptoms from VA Emergency Department and Primary Care Clinical Documentation." *International Journal of Medical Informatics* 81(3):143–56.

Morse, Stephen S. 2012. "Public Health Surveillance and Infectious Disease Detection." *Biosecurity and Bioterrorism: Biodefense Strategy, Practice, and Science* 10(1):6–16.

Pineda, Arturo López, Ye Ye, Shyam Visweswaran, Gregory F. Cooper, Michael M. Wagner, and Fuchiang Rich Tsui. 2015. "Comparison of Machine Learning Classifiers for Influenza Detection from Emergency Department Free-Text Reports." *Journal of Biomedical Informatics* 58:60–69.

Rajput, Nikhil Kumar, Bhavya Ahuja Grover, and Vipin Kumar Rathi. 2020. "Word Frequency and Sentiment Analysis of Twitter Messages during Coronavirus Pandemic." *ArXiv Preprint ArXiv:2004.03925*.

Savova, Guergana K., James J. Masanz, Philip V Ogren, Jiaping Zheng, Sunghwan Sohn, Karin C. Kipper-Schuler, and Christopher G. Chute. 2010. "Mayo Clinical Text Analysis and Knowledge Extraction System (CTAKES): Architecture, Component Evaluation and Applications." *Journal of the American Medical Informatics Association* 17(5):507–13.

Singh, Lisa, Shweta Bansal, Leticia Bode, Ceren Budak, Guangqing Chi, Kornraphop Kawintiranon, Colton Padden, Rebecca Vanarsdall, Emily Vraga, and Yanchen Wang. 2020. "A First Look at COVID-19 Information and Misinformation Sharing on Twitter." *ArXiv Preprint ArXiv:2003.13907*.

Soysal, Ergin, Jingqi Wang, Min Jiang, Yonghui Wu, Serguei Pakhomov, Hongfang Liu, and Hua Xu. 2018. "CLAMP–a Toolkit for Efficiently Building Customized Clinical Natural Language Processing Pipelines." *Journal of the American Medical Informatics Association* 25(3):331–36.

Wang, Lucy Lu, Kyle Lo, Yoganand Chandrasekhar, Russell Reas, Jiangjiang Yang, Darrin Eide, Kathryn Funk, Rodney Kinney, Ziyang Liu, and William Merrill.

2020. "CORD-19: The Covid-19 Open Research Dataset." *ArXiv Preprint ArXiv:2004.10706.*

World Health Organization. 2020a. "Naming the Coronavirus Disease (COVID-19) and the Virus That Causes It." Retrieved June 10, 2020 (https://www.who.int/emergencies/diseases/novel-coronavirus-2019/technical-guidance/naming-the-coronavirus-disease-(covid-2019)-and-the-virus-that-causes-it).

World Health Organization. 2020b. "WHO Director-General's Opening Remarks at the Media Briefing on COVID-19 - 25 May 2020." Retrieved June 10, 2020 (https://www.who.int/dg/speeches/detail/who-director-general-s-opening-remarks-at-the-media-briefing-on-covid-19---25-may-2020).

Xu, Bo, Bernardo Gutierrez, Sumiko Mekaru, Kara Sewalk, Lauren Goodwin, Alyssa Loskill, Emily L. Cohn, Yulin Hswen, Sarah C. Hill, and Maria M. Cobo. 2020. "Epidemiological Data from the COVID-19 Outbreak, Real-Time Case Information." *Scientific Data* 7(1):1–6.

Appendix A: NLP Pipeline

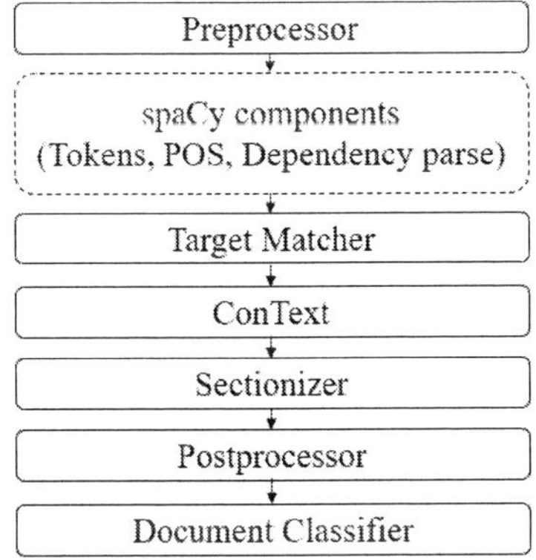

Figure 3. Diagram of components in modular text processing pipeline. Components developed in this work marked by a solid line and existing spaCy components by a dashed line.

Appendix B: Error Analysis

Template or educational text
"Do you have any: * Fever * **Diagnosed with** *COVID-19* in the last 14 days" "The patient reports that they have _ _ _ _ _ **diagnosed with** *COVID-19*"
Experiencer other than the patient
"Veteran's *ex* tested **positive for** *COVID-19.*" "Patient's wife is a nurse. *She* **tested positive** for *coronavirus.*"
Incorrectly linked modifiers
"They said he has not **presented with** any sxs of *COVID-19.*" "Veteran with decreased **positive** *lifestyle* due to *COVID-19.*"
Uncertain
"Admitting Diagnosis: COVID CHECK"
Not relevant to patient diagnosis
"TELEHEALTH SCREENING: Called to explain program **COVID-19** + Monitoring" "**75 yo man with** telephone primary care follow-up due to *COVID-19 restrictions.*"

Table 2. Examples and explanations of false positives.

Text formatting causes incorrect sentence splitting
"Employee was tested for *COVID<END OF SENTENCE>* XX/XX/2020 and result **positive.**"
Positive modifier too far from target concept
"Contacted Veteran for daily follow-up for *COVID-19* screening. Discussed the following: Employee tested **positive.**"
Incorrectly linked modifiers
"**Risk for** respiratory insufficiency r/t *COVID-19.*"
Variations on positive modifiers not recognized by system
"**62 y M** *COVID-19*" (variation of "**62 year old Male with** *COVID-19*")

Table 3. Examples and explanations of false negatives.

Measuring Emotions in the COVID-19 Real World Worry Dataset

Bennett Kleinberg[1,2]   **Isabelle van der Vegt**[1]   **Maximilian Mozes**[1,2,3]

[1]Department of Security and Crime Science
[2]Dawes Centre for Future Crime
[3]Department of Computer Science
University College London

{bennett.kleinberg, isabelle.vandervegt, maximilian.mozes}@ucl.ac.uk

Abstract

The COVID-19 pandemic is having a dramatic impact on societies and economies around the world. With various measures of lockdowns and social distancing in place, it becomes important to understand emotional responses on a large scale. In this paper, we present the first ground truth dataset of emotional responses to COVID-19. We asked participants to indicate their emotions and express these in text. This resulted in the *Real World Worry Dataset* of 5,000 texts (2,500 short + 2,500 long texts). Our analyses suggest that emotional responses correlated with linguistic measures. Topic modeling further revealed that people in the UK worry about their family and the economic situation. Tweet-sized texts functioned as a call for solidarity, while longer texts shed light on worries and concerns. Using predictive modeling approaches, we were able to approximate the emotional responses of participants from text within 14% of their actual value. We encourage others to use the dataset and improve how we can use automated methods to learn about emotional responses and worries about an urgent problem.

1   Introduction

The outbreak of the SARS-CoV-2 virus in late 2019 and subsequent evolution of the COVID-19 disease has affected the world on an enormous scale. While hospitals are at the forefront of trying to mitigate the life-threatening consequences of the disease, practically all societal levels are dealing directly or indirectly with an unprecedented situation. Most countries are — at the time of writing this paper — in various stages of a lockdown. Schools and universities are closed or operate online-only, and merely essential shops are kept open.

At the same time, lockdown measures such as social distancing (e.g., keeping a distance of at least 1.5 meters from one another and only socializing with two people at most) might have a direct impact on people's mental health. With an uncertain outlook on the development of the COVID-19 situation and its preventative measures, it is of vital importance to understand how governments, NGOs, and social organizations can help those who are most affected by the situation. That implies, at the first stage, understanding the emotions, worries, and concerns that people have and possible coping strategies. Since a majority of online communication is recorded in the form of text data, measuring the emotions around COVID-19 will be a central part of understanding and addressing the impacts of the COVID-19 situation on people. This is where computational linguistics can play a crucial role.

In this paper, we present and make publicly available a high quality, ground truth text dataset of emotional responses to COVID-19. We report initial findings on linguistic correlates of emotions, topic models, and prediction experiments.

1.1   Ground truth emotions datasets

Tasks like emotion detection (Seyeditabari et al., 2018) and sentiment analysis (Liu, 2015) typically rely on labeled data in one of two forms. Either a corpus is annotated on a document-level, where individual documents are judged according to a predefined set of emotions (Strapparava and Mihalcea, 2007; Preoţiuc-Pietro et al., 2016) or individual n-grams sourced from a dictionary are categorised or scored with respect to their emotional value (Bradley et al., 1999; Strapparava and Valitutti, 2004). These annotations are done (semi) automatically (e.g., exploiting hashtags such as #happy) (Mohammad and Kiritchenko, 2015; Abdul-Mageed and Ungar, 2017) or manually through third persons (Mohammad and Turney, 2010). While these approaches are common practice and have accelerated the progress that was made in the field, they are limited in that they propagate a *pseudo* ground truth. This is problematic because, as we argue, the core aim of emotion detection is to make an inference about the author's emotional state. The text as the product of an emotional state then functions as a proxy for the latter. For example, rather than wanting to know whether a Tweet is written in a pessimistic tone, we are interested in learning whether the author of the text actually felt pessimistic.

The limitation inherent to third-person annotation, then, is that they might not be adequate measurements of the emotional state of interest. The solution, albeit a costly one, lies in ground truth datasets. Whereas real ground truth would require - in its strictest sense - a random assignment of people to experimental conditions (e.g., one group that is given a positive product experience, and another group with a negative experience), variations that rely on self-reported emotions can also mitigate the problem. A dataset that relies on self-reports is the *International Survey on Emotion Antecedents and Reactions* (ISEAR)[1], which asked participants to recall from memory situations that evoked a set of emotions. The COVID-19 situation is unique and calls for novel datasets that capture people's affective responses to it while it is happening.

1.2 Current COVID-19 datasets

Several datasets mapping how the public responds to the pandemic have been made available. For example, tweets relating to the Coronavirus have been collected since March 11, 2020, yielding about 4.4 million tweets a day (Banda et al., 2020). Tweets were collected through the Twitter stream API, using keywords such as 'coronavirus' and 'COVID-19'. Another Twitter dataset of Coronavirus tweets has been collected since January 22, 2020, in several languages, including English, Spanish, and Indonesian (Chen et al., 2020). Further efforts include the ongoing Pandemic Project[2] which has people write about the effect of the coronavirus outbreak on their everyday lives.

1.3 The COVID-19 Real World Worry Dataset

This paper reports initial findings for the *Real World Worry Dataset* (RWWD) that captured the emotional responses of UK residents to COVID-19

at a point in time where the impact of the COVID-19 situation affected the lives of all individuals in the UK. The data were collected on the 6th and 7th of April 2020, a time at which the UK was under "lockdown" (news, 2020), and death tolls were increasing. On April 6, 5,373 people in the UK had died of the virus, and 51,608 tested positive (Walker , now). On the day before data collection, the Queen addressed the nation via a television broadcast (Guardian, 2020). Furthermore, it was also announced that Prime Minister Boris Johnson was admitted to intensive care in a hospital for COVID-19 symptoms (Lyons, 2020).

The RWWD is a ground truth dataset that used a direct survey method and obtained written accounts of people alongside data of their felt emotions while writing. As such, the dataset does not rely on third-person annotation but can resort to direct self-reported emotions. We present two versions of RWWD, each consisting of 2,500 English texts representing the participants' genuine emotional responses to Corona situation in the UK: the Long RWWD consists of texts that were open-ended in length and asked the participants to express their feelings as they wish. The Short RWWD asked the same people also to express their feelings in Tweet-sized texts. The latter was chosen to facilitate the use of this dataset for Twitter data research.

The dataset is publicly available.[3]

2 Data

We collected the data of $n = 2500$ participants (94.46% native English speakers) via the crowd-sourcing platform Prolific[4]. Every participant provided consent in line with the local IRB. The sample requirements were that the participants were resident in the UK and a Twitter user. In the data collection task, all participants were asked to indicate how they felt about the current COVID-19 situation using 9-point scales (1 = not at all, 5 = moderately, 9 = very much). Specifically, each participant rated how worried they were about the Corona/COVID-19 situation and how much anger, anxiety, desire, disgust, fear, happiness, relaxation, and sadness (Harmon-Jones et al., 2016) they felt about their situation at this moment. They also had to choose which of the eight emotions (ex-

---

[1] https://www.unige.ch/cisa/research/
materials-and-online-research/
research-material/

[2] https://utpsyc.org/covid19/index.html

[3] Data: https://github.com/ben-aaron188/
covid19worry and https://osf.io/awy7r/

[4] https://www.prolific.co/

cept worry) best represented their feeling at this moment.

All participants were then asked to write two texts. First, we instructed them to *"write in a few sentences how you feel about the Corona situation at this very moment. This text should express your feelings at this moment"* (min. 500 characters). The second part asked them to express their feelings in Tweet form (max. 240 characters) with otherwise identical instructions. Finally, the participants indicated on a 9-point scale how well they felt they could express their feelings (in general/in the long text/in the Tweet-length text) and how often they used Twitter (from 1=never, 5=every month, 9=every day) and whether English was their native language. The overall corpus size of the dataset was 2500 long texts (320,372 tokens) and 2500 short texts (69,171 tokens). In long and short texts, only 6 and 17 emoticons (e.g. ":(", "<3") were found, respectively. Because of the low frequency of emoticons, these were not focused on in our analysis.

2.1 Excerpts

Below are two excerpts from the dataset:

Long text: *I am 6 months pregnant, so I feel worried about the impact that getting the virus would have on me and the baby. My husband also has asthma so that is a concern too. I am worried about the impact that the lockdown will have on my ability to access the healthcare I will need when having the baby, and also about the exposure to the virus [...] There is just so much uncertainty about the future and what the coming weeks and months will hold for me and the people I care about.*

Tweet-sized text: *Proud of our NHS and keyworkers who are working on the frontline at the moment. I'm optimistic about the future, IF EVERYONE FOLLOWS THE RULES. We need to unite as a country, by social distancing and stay in.*

2.2 Descriptive statistics

We excluded nine participants who padded the long text with punctuation or letter repetitions. The dominant feelings of participants were anxiety/worry, sadness, and fear (see Table 1)[5]. For all emotions,

the participants' self-rating ranged across the whole spectrum (from "not at all" to "very much"). The final sample consisted to 65.15% of females[6] with an overall mean age of 33.84 years ($SD = 22.04$).

The participants' self-reported ability to express their feelings, in general, was $M = 6.88$ ($SD = 1.69$). When specified for both types of texts separately, we find that the ability to express themselves in the long text ($M = 7.12$, $SD = 1.78$) was higher than that for short texts ($M = 5.91$, $SD = 2.12$), Bayes factor $> 1e + 96$.

The participants reported to use Twitter almost weekly ($M = 6.26$, $SD = 2.80$), tweeted themselves rarely to once per month ($M = 3.67$, $SD = 2.52$), and actively participated in conversations in a similar frequency ($M = 3.41$, $SD = 2.40$). Our participants were thus familiar with Twitter as a platform but not overly active in tweeting themselves.

Variable	Mean	SD
Corpus descriptives		
Tokens (long text)	127.75	39.67
Tokens (short text)	27.70	15.98
Types (long text)	82.69	18.24
Types (short text)	23.50	12.21
TTR (long text)	0.66	0.06
TTR (short text)	0.88	0.09
Chars. (long text)	632.54	197.75
Chars. (short text)	137.21	78.40
Emotions		
Worry	6.55^a	1.76
Anger[1] (4.33%)	3.91^b	2.24
Anxiety (55.36%)	6.49^a	2.28
Desire (1.09%)	2.97^b	2.04
Disgust (0.69%)	3.23^b	2.13
Fear (9.22%)	5.67^a	2.27
Happiness (1.58%)	3.62^b	1.89
Relaxation (13.38%)	3.95^b	2.13
Sadness (14.36%)	5.59^a	2.31

Table 1: Descriptive statistics of text data and emotion ratings. [1]brackets indicate how often the emotion was chosen as the best fit for the current feeling about COVID-19. [a]the value is larger than the neutral midpoint with Bayes factors $> 1e + 32$. [b]the value is smaller than the neutral midpoint with BF $> 1e + 115$. TTR = type-token ratio.

---

[5]For correlations among the emotions, see the online supplement

[6]For an analysis of gender differences using this dataset, see van der Vegt and Kleinberg (2020).

3   Findings and experiments

3.1   Correlations of emotions with LIWC categories

We correlated the self-reported emotions to matching categories of the LIWC2015 lexicon (Pennebaker et al., 2015). The overall matching rate was high (92.36% and 90.11% for short and long texts, respectively). Across all correlations, we see that the extent to which the linguistic variables explain variance in the emotion values (indicated by the R^2) is larger in long texts than in Tweet-sized short texts (see Table 2). There are significant positive correlations for all affective LIWC variables with their corresponding self-reported emotions (i.e., higher LIWC scores accompanied higher emotion scores, and vice versa). These correlations imply that the linguistic variables explain up to 10% and 3% of the variance in the emotion scores for long and short texts, respectively.

The LIWC also contains categories intended to capture areas that concern people (not necessarily in a negative sense), which we correlated to the self-reported worry score. Positive (negative) correlations would suggest that the higher (lower) the worry score of the participants, the larger their score on the respective LIWC category. We found no correlation between the categories "work", "money" and "death" suggesting that the worry people reported was not associated with these categories. Significant positive correlations emerged for long texts for "family" and "friend": the more people were worried, the more they spoke about family and — to a lesser degree — friends.

3.2   Topic models of people's worries

We constructed topic models for both the long and short texts separately using the stm package in R (Roberts et al., 2014a). The text data were lowercased, punctuation, stopwords and numbers were removed, and all words were stemmed. For the long texts, we chose a topic model with 20 topics as determined by semantic coherence and exclusivity values for the model (Mimno et al., 2011; Roberts et al., 2014b,a). Table 3 shows the five most prevalent topics with ten associated frequent terms for each topic (see online supplement for all 20 topics). The most prevalent topic seems to relate to following the rules related to the lockdown. In contrast, the second most prevalent topic appears to relate to worries about employment and the economy. For the Tweet-sized texts, we selected a model with 15 topics. The most common topic bears a resemblance to the government slogan "Stay at home, protect the NHS, save lives." The second most prevalent topic seems to relate to calls for others to adhere to social distancing rules.

3.3   Predicting emotions about COVID-19

It is worth noting that the current literature on automatic emotion detection mainly casts this problem as a classification task, where words or documents are classified into emotional categories (Buechel and Hahn, 2016; Demszky et al., 2020). Our fine-grained annotations allow for estimating emotional values on a continuous scale. Previous works on emotion regression utilise supervised models such as linear regression for this task (Preoţiuc-Pietro et al., 2016), and more recent efforts employ neural network-based methods (Wang et al., 2016; Zhu et al., 2019). However, the latter typically require larger amounts of annotated data, and are hence less applicable to our collected dataset.

We, therefore, use linear regression models to predict the reported emotional values (i.e., anxiety, fear, sadness, worry) based on text properties. Specifically, we applied regularised ridge regression models[7] using TFIDF and part-of-speech (POS) features extracted from long and short texts separately. TFIDF features were computed based on the 1000 most frequent words in the vocabularies of each corpus; POS features were extracted using a predefined scheme of 53 POS tags in *spaCy*[8].

We process the resulting feature representations using principal component analysis and assess the performances using the mean absolute error (MAE) and the coefficient of determination R^2. Each experiment is conducted using five-fold cross-validation, and the arithmetic means of all five folds are reported as the final performance results.

Table 4 shows the performance results in both long and short texts. We observe MAEs ranging between 1.26 (worry with TFIDF) and 1.88 (sadness with POS) for the long texts, and between 1.37 (worry with POS) and 1.91 (sadness with POS) for the short texts. We furthermore observe that the models perform best in predicting the worry scores for both long and short texts. The models explain up to 16% of the variance for the emotional response variables on the long texts, but only up to

---

[7]We used the *scikit-learn* python library (Pedregosa et al., 2011).

[8]https://spacy.io

Correlates	Long texts	Short texts
Affective processes		
Anger - LIWC "anger"	0.28 [0.23; 0.32] (7.56%)	0.09 [0.04; 0.15] (0.88%)
Sadness - LIWC "sad"	0.21 [0.16; 0.26] (4.35%)	0.13 [0.07; 0.18] (1.58%)
Anxiety - LIWC "anx"	0.33 [0.28; 0.37] (10.63%)	0.18 [0.13; 0.23] (3.38%)
Worry - LIWC "anx"	0.30 [0.26; 0.35] (9.27%)	0.18 [0.13; 0.23] (3.30%)
Happiness - LIWC "posemo"	0.22 [0.17; 0.26] (4.64%)	0.13 [0.07; 0.18] (1.56%)
Concern sub-categories		
Worry - LIWC "work"	-0.03 [-0.08; 0.02] (0.01%)	-0.03 [-0.08; 0.02] (0.10%)
Worry - LIWC "money"	0.00 [-0.05; 0.05] (0.00%)	-0.01 [-0.06; 0.04] (0.00%)
Worry - LIWC "death"	0.05 [-0.01; 0.10] (0.26%)	0.05 [0.00; 0.10] (0.29%)
Worry - LIWC "family"	0.18 [0.13; 0.23] (3.12%)	0.06 [0.01; 0.11] (0.40%)
Worry - LIWC "friend"	0.07 [0.01; 0.12] (0.42%)	-0.01 [-0.06; 0.05] (0.00%)

Table 2: Correlations (Pearson's r, 99% CI, R-squared in %) between LIWC variables and emotions.

Docs	Terms
Long texts	
9.52	people, take, think, rule, stay, serious, follow, virus, mani, will
8.35	will, worri, job, long, also, economy, concern, impact, famili, situat
7.59	feel, time, situat, relax, quit, moment, sad, thing, like, also
6.87	feel, will, anxious, know, also, famili, worri, friend, like, sad
5.69	work, home, worri, famili, friend, abl, time, miss, school, children
Short texts	
10.70	stay, home, safe, live, pleas, insid, save, protect, nhs, everyone
8.27	people, need, rule, dont, stop, selfish, social, die, distance, spread
7.96	get, can, just, back, wish, normal, listen, lockdown, follow, sooner
7.34	famili, anxious, worri, scare, friend, see, want, miss, concern, covid
6.81	feel, situat, current, anxious, frustrat, help, also, away, may, extrem

Table 3: The five most prevalent topics for long and short texts.

1% on Tweet-sized texts.

Model	Long		Short	
	MAE	R^2	MAE	R^2
Anxiety - TFIDF	1.65	0.16	1.82	-0.01
Anxiety - POS	1.79	0.04	1.84	0.00
Fear - TFIDF	1.71	0.15	1.85	0.00
Fear - POS	1.83	0.05	1.87	0.01
Sadness - TFIDF	1.75	0.12	1.90	-0.02
Sadness - POS	1.88	0.02	1.91	-0.01
Worry - TFIDF	1.26	0.16	1.38	-0.03
Worry - POS	1.35	0.03	1.37	0.01

Table 4: Results for regression modeling for long and short texts.

4 Discussion

This paper introduced a ground truth dataset of emotional responses in the UK to the Corona pandemic. We reported initial findings on the linguistic correlates of emotional states, used topic modeling to understand what people in the UK are concerned about, and ran prediction experiments to infer emotional states from text using machine learning. These analyses provided several core findings: (1) Some emotional states correlated with word lists made to measure these constructs, (2) longer texts were more useful to identify patterns in language that relate to emotions than shorter texts, (3) Tweet-sized texts served as a means to call for solidarity during lockdown measures while longer texts gave insights to people's worries, and (4) preliminary regression experiments indicate that we can infer from the texts the emotional responses with an absolute error of 1.26 on a 9-point scale (14%).

4.1 Linguistic correlates of emotions and worries

Emotional reactions to the Coronavirus were obtained through self-reported scores. When we used psycholinguistic word lists that measure these emotions, we found weak positive correlations. The lexicon-approach was best at measuring anger, anxiety, and worry and did so better for longer texts than for Tweet-sized texts. That difference is not surprising given that the LIWC was not constructed for micro-blogging and very short documents. In behavioral and cognitive research, small effects (here: a maximum of 10.63% of explained variance) are the rule rather than the exception (Gelman, 2017; Yarkoni and Westfall, 2017). It is essential, however, to interpret them as such: if 10% of the variance in the anxiety score are explained through a linguistic measurement, 90% are not. An explanation for the imperfect correlations - aside from random measurement error - might lie in the inadequate expression of someone's felt emotion in the form of written text. The latter is partly corroborated by even smaller effects for shorter texts, which may have been too short to allow for the expression of one's emotion.

It is also important to look at the overlap in emotions. Correlational follow-up analysis (see online supplement) among the self-reported emotions showed high correlations of worry with fear ($r = 0.70$) and anxiety ($r = 0.66$) suggesting that these are not clearly separate constructs in our dataset. Other high correlations were evident between anger and disgust ($r = 0.67$), fear and anxiety ($r = 0.78$), and happiness and relaxation ($r = 0.68$). Although the chosen emotions (with our addition of "worry") were adopted from previous work (Harmon-Jones et al., 2016), it merits attention in future work to disentangle the emotions and assess, for example, common ngrams per cluster of emotions (e.g. as in Demszky et al., 2020).

4.2 Topics of people's worries

Prevalent topics in our corpus showed that people worry about their jobs and the economy, as well as their friends and family - the latter of which is also corroborated by the LIWC analysis. For example, people discussed the potential impact of the situation on their family, as well as their children missing school. Participants also discussed the lockdown and social distancing measures. In the Tweet-sized texts, in particular, people encouraged others to stay at home and adhere to lockdown rules in order to slow the spread of the virus, save lives and/or protect the NHS. Thus, people used the shorter texts as a means to call for solidarity, while longer texts offered insights into their actual worries (for recent work on gender differences, see van der Vegt and Kleinberg, 2020).

While there are various ways to select the ideal number of topics, we have relied on assessing the semantic coherence of topics and exclusivity of topic words. Since there does not seem to be a consensus on the best practice for selecting topic numbers, we encourage others to examine different approaches or models with varying numbers of topics.

4.3 Predicting emotional responses

Prediction experiments revealed that ridge regression models can be used to approximate emotional responses to COVID-19 based on encoding of the textual features extracted from the participants' statements. Similar to the correlational and topic modeling findings, there is a stark difference between the long and short texts: the regression models are more accurate and explain more variance for longer than for shorter texts. Additional experiments are required to investigate further the expressiveness of the collected textual statements for the prediction of emotional values. The best predictions were obtained for the reported worry score ($MAE = 1.26$, $MAPE = 14.00\%$). An explanation why worry was the easiest to predict could be that it was the highest reported emotion overall with the lowest standard deviation, thus potentially biasing the model. More fine-grained prediction analyses out of the scope of this initial paper could further examine this.

4.4 Suggestions for future research

The current analysis leaves several research questions untouched. First, to mitigate the limitations of lexicon-approaches, future work on inferring emotions around COVID-19 could expand on the prediction approach (e.g., using different feature sets and models). Carefully validated models could help to provide the basis for large scale, real-time measurements of emotional responses. Of particular importance is a solution to the problem hinted at in the current paper: the shorter, Tweet-sized texts contained much less information, had a different function, and were less suitable for predictive modeling. However, it must be noted that the experimental setup of this study did not fully mimic a 'natural' Twitter experience. Whether the results are generalisable to actual Twitter data is an important empirical question for follow-up work. Nevertheless, with much of today's stream of text data coming in the form of (very) short messages, it is important to understand the limitations of using that kind of data and worthwhile examining how we can better make inferences from that information.

Second, with a lot of research attention paid to readily available Twitter data, we hope that future studies also focus on non-Twitter data to capture emotional responses of those who are underrepresented (or non-represented) on social media but are at heightened risk.

Third, future research may focus on manually annotating topics to more precisely map out what people worry about with regards to COVID-19. Several raters could assess frequent terms for each topic, then assign a label. Then through discussion or majority votes, final topic labels can be assigned to obtain a model of COVID-19 real-world worries.

Fourth, future efforts may aim for sampling over a longer period to capture how emotional responses develop over time. Ideally, using high-frequency sampling (e.g., daily for several months), future work could account for the large number of events that may affect emotions.

Lastly, it is worthwhile to utilise other approaches to measuring psychological constructs in text. Although the rate of out-of-vocabulary terms for the LIWC in our data was low, other dictionaries may be able to capture other relevant constructs. For instance, the tool Empath (Fast et al., 2016) could help measure emotions not available in the LIWC (e.g., nervousness and optimism). We hope that future work will use the current dataset (and extensions thereof) to go further so we can better understand emotional responses in the real world.

5 Conclusions

This paper introduced the first ground truth dataset of emotional responses to COVID-19 in text form. Our findings highlight the potential of inferring concerns and worries from text data but also show some of the pitfalls, in particular, when using concise texts as data. We encourage the research community to use the dataset so we can better understand the impact of the pandemic on people's lives.

Acknowledgments

This research was supported by the Dawes Centre for Future Crime at UCL.

References

Muhammad Abdul-Mageed and Lyle Ungar. 2017. EmoNet: Fine-grained emotion detection with gated recurrent neural networks. In *Proceedings of the 55th Annual Meeting of the Association for Computational Linguistics (Volume 1: Long Papers)*, pages 718–728, Vancouver, Canada. Association for Computational Linguistics.

Juan M. Banda, Ramya Tekumalla, Guanyu Wang, Jingyuan Yu, Tuo Liu, Yuning Ding, and Gerardo Chowell. 2020. A Twitter Dataset of 150+ million tweets related to COVID-19 for open research. Type: dataset.

Margaret M. Bradley, Peter J. Lang, Margaret M. Bradley, and Peter J. Lang. 1999. Affective norms for english words (anew): Instruction manual and affective ratings.

Sven Buechel and Udo Hahn. 2016. Emotion analysis as a regression problem — dimensional models and their implications on emotion representation and metrical evaluation. In *Proceedings of the Twenty-Second European Conference on Artificial Intelligence*, ECAI'16, page 1114–1122, NLD. IOS Press.

Emily Chen, Kristina Lerman, and Emilio Ferrara. 2020. #COVID-19: The First Public Coronavirus Twitter Dataset. Original-date: 2020-03-15T17:32:03Z.

Dorottya Demszky, Dana Movshovitz-Attias, Jeongwoo Ko, Alan Cowen, Gaurav Nemade, and Sujith Ravi. 2020. GoEmotions: A Dataset of Fine-Grained Emotions. *arXiv:2005.00547 [cs]*. ArXiv: 2005.00547.

Ethan Fast, Binbin Chen, and Michael S. Bernstein. 2016. Empath: Understanding Topic Signals in Large-Scale Text. In *Proceedings of the 2016 CHI Conference on Human Factors in Computing Systems*, pages 4647–4657, San Jose California USA. ACM.

Andrew Gelman. 2017. The piranha problem in social psychology / behavioral economics: The "take a pill" model of science eats itself - Statistical Modeling, Causal Inference, and Social Science.

The Guardian. 2020. Coronavirus latest: 5 April at a glance. *The Guardian*.

Cindy Harmon-Jones, Brock Bastian, and Eddie Harmon-Jones. 2016. The Discrete Emotions Questionnaire: A New Tool for Measuring State Self-Reported Emotions. *PLOS ONE*, 11(8):e0159915.

Bing Liu. 2015. *Sentiment analysis: mining opinions, sentiments, and emotions*. Cambridge University Press, New York, NY.

Kate Lyons. 2020. Coronavirus latest: at a glance. *The Guardian*.

David Mimno, Hanna Wallach, Edmund Talley, Miriam Leenders, and Andrew McCallum. 2011. Optimizing Semantic Coherence in Topic Models. page 11.

Saif Mohammad and Peter Turney. 2010. Emotions Evoked by Common Words and Phrases: Using Mechanical Turk to Create an Emotion Lexicon. In *Proceedings of the NAACL HLT 2010 Workshop on Computational Approaches to Analysis and Generation of Emotion in Text*, pages 26–34, Los Angeles, CA. Association for Computational Linguistics.

Saif M. Mohammad and Svetlana Kiritchenko. 2015. Using Hashtags to Capture Fine Emotion Categories from Tweets. *Computational Intelligence*, 31(2):301–326.

ITV news. 2020. Police can issue 'unlimited fines' to those flouting coronavirus social distancing rules, says Health Secretary. Library Catalog: www.itv.com.

F. Pedregosa, G. Varoquaux, A. Gramfort, V. Michel, B. Thirion, O. Grisel, M. Blondel, P. Prettenhofer, R. Weiss, V. Dubourg, J. Vanderplas, A. Passos, D. Cournapeau, M. Brucher, M. Perrot, and E. Duchesnay. 2011. Scikit-learn: Machine learning in Python. *Journal of Machine Learning Research*, 12:2825–2830.

James W. Pennebaker, Ryan L. Boyd, Kayla Jordan, and Kate Blackburn. 2015. The development and psychometric properties of LIWC2015. Technical report.

Daniel Preoţiuc-Pietro, H. Andrew Schwartz, Gregory Park, Johannes Eichstaedt, Margaret Kern, Lyle Ungar, and Elisabeth Shulman. 2016. Modelling valence and arousal in Facebook posts. In *Proceedings of the 7th Workshop on Computational Approaches to Subjectivity, Sentiment and Social Media Analysis*, pages 9–15, San Diego, California. Association for Computational Linguistics.

Margaret E Roberts, Brandon M Stewart, and Dustin Tingley. 2014a. stm: R Package for Structural Topic Models. *Journal of Statistical Software*, page 41.

Margaret E. Roberts, Brandon M. Stewart, Dustin Tingley, Christopher Lucas, Jetson Leder-Luis, Shana Kushner Gadarian, Bethany Albertson, and David G. Rand. 2014b. Structural Topic Models

for Open-Ended Survey Responses. *American Journal of Political Science*, 58(4):1064–1082. _eprint: https://onlinelibrary.wiley.com/doi/pdf/10.1111/ajps.12103.

Armin Seyeditabari, Narges Tabari, and Wlodek Zadrozny. 2018. Emotion Detection in Text: a Review. *arXiv:1806.00674 [cs]*. ArXiv: 1806.00674.

Carlo Strapparava and Rada Mihalcea. 2007. SemEval-2007 task 14: Affective text. In *Proceedings of the Fourth International Workshop on Semantic Evaluations (SemEval-2007)*, pages 70–74, Prague, Czech Republic. Association for Computational Linguistics.

Carlo Strapparava and Alessandro Valitutti. 2004. WordNet affect: an affective extension of WordNet. In *Proceedings of the Fourth International Conference on Language Resources and Evaluation (LREC'04)*, Lisbon, Portugal. European Language Resources Association (ELRA).

Isabelle van der Vegt and Bennett Kleinberg. 2020. Women worry about family, men about the economy: Gender differences in emotional responses to COVID-19. *arXiv:2004.08202 [cs]*. ArXiv: 2004.08202.

Amy Walker (now), Matthew Weaver (earlier), Steven Morris, Jamie Grierson, Mark Brown, Jamie Grierson, and Pete Pattisson. 2020. UK coronavirus live: Boris Johnson remains in hospital 'for observation' after 'comfortable night'. *The Guardian*.

Jin Wang, Liang-Chih Yu, K. Robert Lai, and Xuejie Zhang. 2016. Dimensional sentiment analysis using a regional CNN-LSTM model. In *Proceedings of the 54th Annual Meeting of the Association for Computational Linguistics (Volume 2: Short Papers)*, pages 225–230, Berlin, Germany. Association for Computational Linguistics.

Tal Yarkoni and Jacob Westfall. 2017. Choosing Prediction Over Explanation in Psychology: Lessons From Machine Learning. *Perspectives on Psychological Science*, 12(6):1100–1122.

Suyang Zhu, Shoushan Li, and Guodong Zhou. 2019. Adversarial attention modeling for multidimensional emotion regression. In *Proceedings of the 57th Annual Meeting of the Association for Computational Linguistics*, pages 471–480, Florence, Italy. Association for Computational Linguistics.

Estimating the effect of COVID-19 on mental health:
Linguistic indicators of depression during a global pandemic

JT Wolohan
Booz Allen Hamilton
`wolohan_john@bah.com`

Abstract

This preliminary analysis uses a deep LSTM neural network with fastText embeddings to predict population rates of depression on Reddit in order to estimate the effect of COVID-19 on mental health. We find that year over year, depression rates on Reddit are up 50% , suggesting a 15-million person increase in the number of depressed Americans and a \$7.5 billion increase in depression related spending. This finding comes at a time when uncertainty about the impact of COVID-19 on physical and economic health is still high, and suggests that in addition to those factors, mental health must be considered as well. As data becomes available, further research will be needed to validate the results of this preliminary investigation.

1 Introduction

The COVID-19 pandemic has already plagued the people of the world's physical health, and while the impact of the novel coronavirus on our mental health is less well understood, it is expected to be negative (Wang et al., 2020; Ammerman et al., 2020). Even those who never get sick during a pandemic can experience a multitude of psychological stressors during a disease outbreak and those stressors can persist well past the end of the outbreak (Chew et al., 2020). The popular media is aware of the necessity for otherwise healthy people to emphasize "self care"—small acts intended to maintain one's mental health or relieve stress—during these uncertain times. This analysis attempts to quantify the impact that COVID-19 has had to date on the population rate of depression through the use of state-of-science depression prediction models and data from the popular social media site Reddit.

This paper continues an established line of research in the application of natural language processing techniques to the disease of depression (Guntuku et al., 2017), and mixes it with the rapidly emerging field of COVID-19 research (McKibbin and Fernando, 2020; Duan and Zhu, 2020). Research in the former area is centered around the notion that language use reflects the thought processes of the speaker and that by assessing the words people use, we can gain insight into their thought processes (Fine, 2006). From this, it follows that text classification approaches such as the use of long short-term memory networks (Hochreiter and Schmidhuber, 1997) and word embeddings (Mikolov et al., 2013), such as fastText (Bojanowski et al., 2017), can be used to classify people's mental health status based on their speech. Depression has been studied widely in this way due to its grave impact on those it afflicts (De Choudhury et al., 2013; Coppersmith et al., 2015). Indeed, even subclinical levels of depression have been shown to reduce quality of life in meaningful and measurable ways (Cuijpers and Smit, 2002).

2 Method

In this paper, we use data from Reddit to measure the potential impact of COVID-19 on depression. We do this in two parts: first, we extend the approach from Wolohan et al. (2018), using an LSTM model to improve the accuracy of depression prediction on social media; then, with this model, we analyze a new dataset—composed of the Reddit comments of 20,000 users across the first six months of 2018, 2019, and 2020[1]—in order to estimate the population rate of depression during the COVID-19 pandemic.

2.1 Data

For these analyses we use two datasets of Reddit comments, aggregated at the user level. The first dataset–the Off-Topic Depression dataset–comes from Wolohan et al.; The second, is a novel dataset

---

[1] This paper was written in April, 2020. More data will be gathered as it becomes available.

created for this task: the Reddit Pandemic Depression dataset.

The Off-Topic Depression dataset contains 141-million words from Reddit comments, aggregated by 11,000 users. The text in this dataset was not allowed to come from subreddits—the site-wide term for a themed message boards—where discussion of depression or a related issues was expected. Therefore, this dataset contains only "off-topic" text—text not on the subject of depression. This step is important for being able to detect depression among the general public, many of whom are reluctant to talk about their depressive symptoms due to depression-related stigma (Manos et al., 2009). We label users as either "depressed" or "not-depressed" based on self-disclosure behavior: authoring posts in depression-related subeditors. The Off-Topic Depression dataset is useful for training a model because of the available baseline, but contains no data from the recent COVID-19 pandemic. This makes it insufficient to estimate the impact of COVID-19 on population rates of depression.

2.1.1 Pandemic Depression dataset

The Pandemic Depression dataset contains 23 million words generated by 20,000 users over three years. The users for this dataset were selected in a similar fashion to the users from the Off-Topic Depression dataset: a scrape of submission authors to the subreddit r/AskRedit was performed to gather a set of potential users and then a random subset from this set was taken. During the scrape, 29-million users were considered. From that sample, 20,000 were randomly selected for inclusion in the study. This approach attempts to gather "neutral" Reddit users, who are not necessarily associated with any particular subreddit or community of subreddits, and would therefore have no bias towards or away from depression. The data for the Pandemic Depression dataset is broken up by the time of users activities. We use only the first six months of 2018 and 2019 and the first four months of 2020[2].

2.2 Deep LSTM with fastText

As part of this analysis we trained a deep long short-term memory neural network. The network we used contains five layers: a fastText (Joulin

Month	2018	2019	2020
Jan	1.6 mil	5.5 mil.	560k
Feb	475k	1.8 mil.	860k
Mar	335k	1.2 mil	2.4 mil
Apr	258k	1 mil	5.6 mil
May	211k	700k	*
Jun	210k	470k	*
* Data not yet available.			

Table 1: Pandemic Depression dataset text by month.

et al., 2016) embedding layer, three LSTM layers, and an output layer. We trained and evaluated the LSTM on the Off-Topic Depression dataset, using about 7,700 users for training and about 4,000 users for testing. This totalled about 100 million words for training and 40 million for testing.

The first layer of the model, the embedding layer learns weights that take advantage of the specific 300-dimensional fastText vectors. The second through fourth layers of the model are identical LSTM layers with a 20% dropout rate. The LSTM layers use each word as a step. The fifth and final layer of the model is a single-node dense layer with sigmoid activation used for predicting the class of the user. A depiction of this network and the minimal prepossessing can be seen in Figure 1.

Text preprocessing for the LSTM was minimal. The vocabulary for all documents was limited to the most-common 10,000 words. Each user was truncated down to or zero-padded up to 750 words, as necessary. We performed no other preprocessing, such as misspelling correction or internet-speak normalization.

2.3 Comparative time-series analysis

In order to assess the impact of the COVID-19 pandemic on language use, we perform a comparative time-series analysis of three periods: two six month periods from before the pandemic ranging from January 2018 and 2019 to June 2018 and 2019 inclusive, and one four month period from January 2020 to April 2020[3]. We analyzed the same users for all periods. Rates of depression were estimated for each period by classifying each user as either depressed or not-depressed with the LSTM classifier described in section 2.2.

It is important to note here that Reddit activity is inconsistent and that, unlike other social media

---

[2]Historical Gaffney and Matias (2018) note that there are issues with historical analysis of Reddit data—perhaps inclusive of users deleting depressive content post-hoc. Addressing those concerns are outside the scope of this preliminary investigation.

[3]Data from April was only available up to April 4; however data is included because a large enough volume of data was accessible: $\approx 50,000$ words.

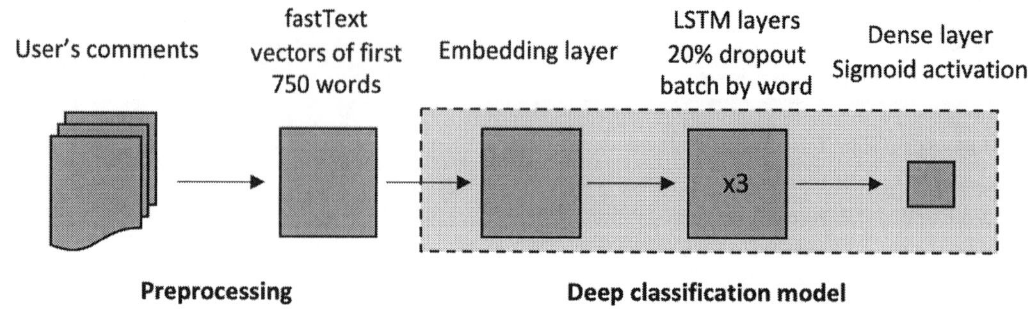

Figure 1: Deep LSTM model for depression prediction.

platforms such as Twitter or Facebook, the use of a single account through time is not encouraged by the platform (Leavitt, 2015). This results in many accounts being abandoned and the resulting 2020 subset of data being smaller in terms of total active users than the subset of data for 2019. A more appropriate means of performing this analysis would be to select a random sample users known to be active during this time period in 2020. Requiring the same users for all periods may introduce bias if users who are likely to be active over long periods of time have a bias towards or away from depression.

3 Preliminary results

In this section, we review the preliminary results. We find that an LSTM with fastText embeddings outperforms the baseline approach in Wolohan et al. Additionally, the LSTM indicates that the population rate of depression may be up by 50% in the first four months of 2020 when compared to the first four months of 2019 and 2018.

3.1 LSTM with fastText embeddings

Comparing the new model for off-topic depression prediction, a deep LSTM with fastText word embeddings, to the model previous used by Wolohan et al., We find that the LSTM outperforms the previous baseline approach in the relevant measures of AUC and F1 score. The LSTM achieved an AUC of 0.93 and an F1 score of 0.92, surpassing the baseline by 18 points 24 points respectively. The results for this LSTM are competitive with state-of-the-art deep-learning approaches for this task on similar datasets (see: Orabi et al. 2018; Guntuku et al. 2017).

Model	AUC	F1
Wolohan et al.	0.75	0.68
LSTM + fastText	0.93	0.92

Table 2: LSTM performance versus baseline.

3.2 Comparative time-series analysis

With the LSTM model improved to 0.93 AUC, we then applied the LSTM to the Pandemic Depression dataset. In doing this, we assessed whether or not users' language indicated depression for the first six months in 2018 and 2019, and the first four months of 2020. We found for 2018 and 2019, the user population being studied demonstrated a steady rate of depression around 33%±4%. For 2020, we found the population rate of depression to average 49%, with individual months ranging from 42% to 52%.

In the first six months of 2018 and 2019, the LSTM suggested a depression rate in the low 30% range with only two exceptions: May 2018 and January 2019. May 2018 had the highest estimated depression rate–38% of all users–while January 2019 had the lowest–29% of all users .

In 2020, estimated depression rates among Reddit users again are consistent; however, they are consistently 20 percentage points higher than 2018 and 2019. Of the three months, only April 2020 stands out with a low depression rate: 42%. At the time of this writing, the data for April 2020 is incomplete.

4 Discussion

With the global COVID-19 pandemic wrecking havoc on medical systems and economies worldwide, 2020 will be a year in which many people go through significant hardships. If the analysis herein is to believed, then the fear that many have about a

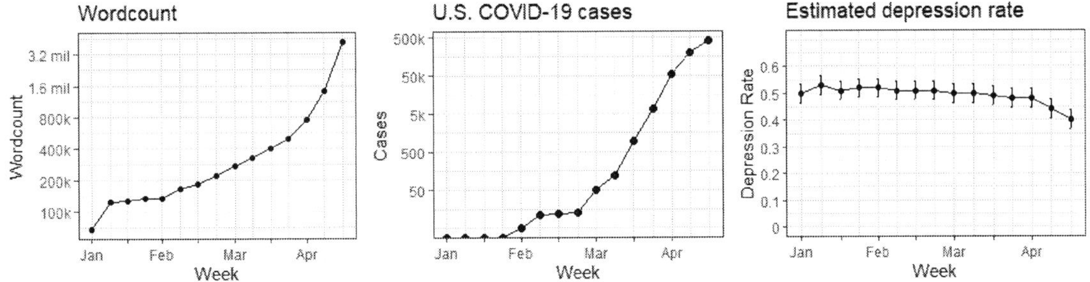

Figure 2: Sampled word count, COVID-19 cases, and modeled depression rate by week.

Month	2018	2019	2020	Δ
January	.32	.29	.52	+79%
February	.34	.32	.51	+69%
March	.31	.32	.49	+53%
April	.33	.34	.42	+24%
May	.38	.32	-	-
June	.34	.34	-	-
Average	.34	.32	.49	+53%

Table 3: Estimated depression rate of Reddit users for select months.

global deterioration in mental health is likely to be a reality as well. But should we believe the analysis presented here; and what are the implications if we do?

4.1 Model efficacy

There are two reasons that we should tentatively believe the AI-based assessments of population-level depression on Reddit. First, the 32% population-rate of depression estimated by the model is plausible, given (1) that the LSTM is designed to detect both clinical and subclinical depression and (2) that Reddit has a much younger (read: depression prone) population than the U.S. at-large. Second, the steadiness of the numbers over time is encouraging.

First among the reasons that one would avoid dismissing these findings out of hand is the consistency between the numbers projected for 2018 and 2019 and what one would expect for a population joint rate of clinical and subclinical depression on Reddit. According to the U.S. National Institute of Health, adult rates of depression range up to 13% depending on the demographic[4]. At the upper bound of 13% is the 18-25 year-old demographic.

More than half of the Reddit-using population falls into this demographic[5]. Further, 25% of Reddit users are 17 or younger—suggesting they might also have an increased rate of depression. NIH estimates that approximately 17% of adolescents between 15 and 17 will experience at least one major depressive episode each year. If one takes 13% as the population rate of depression for Reddit users and doubles it to include cases of subclinical depression (Kessler et al., 1997), one is left with a 26% rate of total depression for Reddit users for the first six months 2019. The LSTM predicts a 32% rate of depression for the first six months of 2019. A reasonable amount of error, given the accuracy measures in Table 2.

Second, the steadiness of the rates of depression in the two control years is encouraging. There is little variation across the first six months of both 2018 and 2019. With only 9 percentage points separating the month with the greatest estimated rate of depression, May '18, and the lowest month, Jan. '19.

4.2 Estimating the effect of the COVID-19 pandemic on mental health

If we assume that the rates of population-wide clinical and subclinical depression are to increase by 50% in 2020, either as a result of the COVID-19 pandemic or otherwise, then we would expect to see a population rate of depression increase from 7% to 10%, and a similar sized increase sub-clinical depression. In the U.S., this would amount to 15 million more adults suffering from clinical depression. This increased rate of depression would amount to a $7.5 billion increase in healthcare spending, assuming $500 per person per year (Kleine-Budde et al., 2013). This assumes that

---

[4]All population rates of depression come from the NIH: `https://www.nimh.nih.gov/health/statistics/major-depression.shtml`

[5]There are two sources for the demographics of Reddit, both from 2016: Barthel et al. (2016) or /u/HurricaneXriks (2016). The latter is used here.

the average rate of depression is about as severe as it is currently.

Importantly, there reasons we may believe this increased rate is not yet associated with COVID-19. As we can see in Figure 2, the estimated depression rate appears to be declining in April at a time when cases and deaths in the U.S. are rising. Speculatively, this may be associated with the phase of the pandemic the U.S.—and therefore most Reddit users—are currently experiencing. Many of these users will be under stay-at-home orders; however, the full toll of the pandemic, including economic destruction and loss of life has not yet been felt. Alternatively, we must consider that the active population of Reddit may be changing as stay-at-home orders and unemployment furnish people with addition free time to use the internet.

One would expect that depression rates increase as stressors such as unemployment, loneliness, and loss of loved ones begin to impact the population at large. If the population rate of depression is elevated for a reason un-related to COVID-19, that could spell even further trouble for Americans' mental health.

5   Ethical considerations

As with all works related to public health and mental health, it is important that we consider the ethical implications of our research and make explicit the ethical justification for the work. In particular, work of this kind–namely, public health surveillance–requires special attention because, while disease surveillance is foundational to good public health practice (Fairchild et al., 2007), it also forgoes the notion of informed consent. When discussing a health issue rife with stigma such as depression, the concern about a researcher mismanaging data and revealing public health information of individuals without their consent is amplified.

Klingler et al. (2017) enumerate eight broad categories of ethical arguments by which researchers justify forgoing the traditional informed consent requirement for conducting public health surveillance. Of those, we argue that our work satisfies the effectiveness, necessity, proportionality, and least infringement requirements. First, with respect to the effectiveness, this study is the first–to my knowledge–quantitative estimate of the impact of COVID-19 on population rate of depression. That makes this data valuable to the public health and mental health communities. Second, with respect

to necessity, it is noteworthy that this approach is minimally intrusive, requiring no contact with the individuals whose comments are used and no interruption of their usage of the Reddit service. For proportionality, third, it is important to note that no individual user-level data was shared at any time during this research, and that the identities of Reddit users are hidden behind pseudonyms, offering them an additional layer of protection. Fourth and finally, the work considered carefully the notion of least infringement, collecting only data that would be necessary for the analysis herein.

Additionally, in a departure from traditional practices in the NLP community, the data underlying this work will only be shared with researchers who both (1) provide a research design or other public-health justification for the use of the data and (2) agree to take the necessary efforts to secure the data.

Ultimately, we view this work as ethically justified based on the precautions noted above and the potentially large increase in population-level depression against which this research warns. If depression is, as we predict, to impact 15 million more Americans through 2020 than in previous years, advanced warning is valuable to the American mental-health system.

6   Conclusion

In this paper we show the effectiveness of an LSTM text classifier using fastText word embeddings at predicting user-level depression and use that classifier to estimate the population rate of depression in April 2020 in the midst of the COVID-19 pandemic. We estimate that through the first six months of 2020, population rate of depression is up $\approx 50\%$, corresponding to a 15 million more depressed Americans. This analysis suffers from a lack of data and will strengthen as more data becomes available. Additional research is needed to confirm or contradict the results presented here, and will be especially valuable when the adjusted population rates of depression are known.

References

Brooke A Ammerman, Taylor A Burke, Ross Jacobucci, and Kenneth McClure. 2020. Preliminary investigation of the association between covid-19 and suicidal thoughts and behaviors in the u.s.

Michael Barthel, Galen Stocking, Jesse Holcomb, and Amy Mitchell. 2016. Reddit news users more likely to be male, young and digital in their news preferences. shorturl.at/chsG8. Accessed: 2020-04-07.

Piotr Bojanowski, Edouard Grave, Armand Joulin, and Tomas Mikolov. 2017. Enriching word vectors with subword information. *Transactions of the Association for Computational Linguistics*, 5:135–146.

Qian Hui Chew, Ker Chiah Wei, Shawn Vasoo, and Hong Choon Chua. 2020. Narrative synthesis of psychological and coping responses towards emerging infectious disease outbreaks in the general population: practical considerations for the covid-19 pandemic. *Singapore medical journal.*

Glen Coppersmith, Mark Dredze, Craig Harman, Kristy Hollingshead, and Margaret Mitchell. 2015. Clpsych 2015 shared task: Depression and ptsd on twitter. In *Proceedings of the 2nd Workshop on Computational Linguistics and Clinical Psychology: From Linguistic Signal to Clinical Reality*, pages 31–39.

Pim Cuijpers and Filip Smit. 2002. Excess mortality in depression: a meta-analysis of community studies. *Journal of affective disorders*, 72(3):227–236.

Munmun De Choudhury, Michael Gamon, Scott Counts, and Eric Horvitz. 2013. Predicting depression via social media. In *Seventh international AAAI conference on weblogs and social media.*

Li Duan and Gang Zhu. 2020. Psychological interventions for people affected by the covid-19 epidemic. *The Lancet Psychiatry*, 7(4):300–302.

Amy L Fairchild, Ronald Bayer, James Colgrove, and Daniel Wolfe. 2007. *Searching eyes: privacy, the state, and disease surveillance in America*, volume 18. Univ of California Press.

Jonathan Fine. 2006. *Language in psychiatry: A handbook of clinical practice*. Equinox London.

Devin Gaffney and J Nathan Matias. 2018. Caveat emptor, computational social science: Large-scale missing data in a widely-published reddit corpus. *PloS one*, 13(7).

Sharath Chandra Guntuku, David B Yaden, Margaret L Kern, Lyle H Ungar, and Johannes C Eichstaedt. 2017. Detecting depression and mental illness on social media: an integrative review. *Current Opinion in Behavioral Sciences*, 18:43–49.

Sepp Hochreiter and Jürgen Schmidhuber. 1997. Long short-term memory. *Neural computation*, 9(8):1735–1780.

Armand Joulin, Edouard Grave, Piotr Bojanowski, and Tomas Mikolov. 2016. Bag of tricks for efficient text classification. *arXiv preprint arXiv:1607.01759.*

Ronald C Kessler, Shanyang Zhao, Dan G Blazer, and Marvin Swartz. 1997. Prevalence, correlates, and course of minor depression and major depression in the national comorbidity survey. *Journal of affective disorders*, 45(1-2):19–30.

Katja Kleine-Budde, Romina Müller, Wolfram Kawohl, Anke Bramesfeld, Jörn Moock, and Wulf Rössler. 2013. The cost of depression–a cost analysis from a large database. *Journal of affective disorders*, 147(1-3):137–143.

Corinna Klingler, Diego Steven Silva, Christopher Schuermann, Andreas Alois Reis, Abha Saxena, and Daniel Strech. 2017. Ethical issues in public health surveillance: a systematic qualitative review. *BMC Public Health*, 17(1):295.

Alex Leavitt. 2015. " this is a throwaway account" temporary technical identities and perceptions of anonymity in a massive online community. In *Proceedings of the 18th ACM Conference on Computer Supported Cooperative Work & Social Computing*, pages 317–327.

Rachel C Manos, Laura C Rusch, Jonathan W Kanter, and Lisa M Clifford. 2009. Depression self-stigma as a mediator of the relationship between depression severity and avoidance. *Journal of Social and Clinical Psychology*, 28(9):1128–1143.

Warwick J McKibbin and Roshen Fernando. 2020. The global macroeconomic impacts of covid-19: Seven scenarios.

Tomas Mikolov, Ilya Sutskever, Kai Chen, Greg S Corrado, and Jeff Dean. 2013. Distributed representations of words and phrases and their compositionality. In *Advances in neural information processing systems*, pages 3111–3119.

Ahmed Husseini Orabi, Prasadith Buddhitha, Mahmoud Husseini Orabi, and Diana Inkpen. 2018. Deep learning for depression detection of twitter users. In *Proceedings of the Fifth Workshop on Computational Linguistics and Clinical Psychology: From Keyboard to Clinic*, pages 88–97.

/u/HurricaneXriks. 2016. Results of the reddit demographics survey 2016. https://imgur.com/gallery/cPzlB. Accessed: 2020-04-07.

Cuiyan Wang, Riyu Pan, Xiaoyang Wan, Yilin Tan, Linkang Xu, Cyrus S Ho, and Roger C Ho. 2020. Immediate psychological responses and associated factors during the initial stage of the 2019 coronavirus disease (covid-19) epidemic among the general population in china. *International Journal of Environmental Research and Public Health*, 17(5):1729.

JT Wolohan, Misato Hiraga, Atreyee Mukherjee, Zeeshan Ali Sayyed, and Matthew Millard. 2018. Detecting linguistic traces of depression in topic-restricted text: attending to self-stigmatized depression with nlp. In *Proceedings of the First International Workshop on Language Cognition and Computational Models*, pages 11–21.

Exploration of Gender Differences in COVID-19 Discourse on Reddit

Jai Aggarwal **Ella Rabinovich** **Suzanne Stevenson**

Department of Computer Science, University of Toronto

`{jai,ella,suzanne}@cs.toronto.edu`

Abstract

Decades of research on differences in the language of men and women have established postulates about preferences in lexical, topical, and emotional expression between the two genders, along with their sociological underpinnings. Using a novel dataset of male and female linguistic productions collected from the Reddit discussion platform, we further confirm existing assumptions about gender-linked affective distinctions, and demonstrate that these distinctions are amplified in social media postings involving emotionally-charged discourse related to COVID-19. Our analysis also confirms considerable differences in topical preferences between male and female authors in spontaneous pandemic-related discussions.

1 Introduction

Research on gender differences in language has a long history spanning psychology, gender studies, sociolinguistics, and, more recently, computational linguistics. A considerable body of linguistic studies highlights the differences between the language of men and women in topical, lexical, and syntactic aspects (Lakoff, 1973; Labov, 1990), and such differences have proven to be accurately detectable by automatic classification tools (Koppel et al., 2002; Schler et al., 2006; Schwartz et al., 2013). Here, we study the differences in male (M) and female (F) language in discussions of COVID-19[1] on the Reddit[2] discussion platform. Responses to the virus on social media have been heavily emotionally-charged, accompanied by feelings of anxiety, grief, and fear, and have discussed far-ranging concerns regarding personal and public health, the economy, and social aspects of life. In this work, we explore how established emotional and topical cross-gender differences are carried over into this pandemic-related discourse. Insights regrading these distinctions will advance our understanding of gender-linked linguistic traits, and may further help to inform public policy and communications around the pandemic.

Research has considered the emotional content of social media on the topic of the COVID pandemic (e.g., Lwin et al., 2020; Stella et al., 2020), but little work has looked specifically at the impact of gender on affective expression (van der Vegt and Kleinberg, 2020). Gender-linked linguistic distinctions across emotional dimensions have been a subject of prolific research (Burriss et al., 2007; Hoffman, 2008; Thelwall et al., 2010), with findings suggesting that women are more likely than men to express positive emotions, while men exhibit higher tendency to dominance, engagement, and control (although see Park et al. (2016) for an alternative finding). van der Vegt and Kleinberg (2020) compared the self-reported emotional state of male vs. female crowdsourced workers who contributed to the Real World Worry Dataset (RWWD, Kleinberg et al., in press), in which they were also asked to write about their feelings around COVID. However, because van der Vegt and Kleinberg (2020) restricted the affective analysis to the workers' emotional ratings, it remains an open question regarding whether, and how, the natural linguistic productions of males and females about COVID will exhibit detectably different patterns of emotion.

Topical analysis of social media during the pandemic has also been a focus of recent work (e.g., Liu et al., 2020; Abd-Alrazaq et al., 2020), again with few studies devoted to gender differences (Thelwall and Thelwall, 2020; van der Vegt and Kleinberg, 2020). Much prior work has found distinctions in topical preferences in spontaneous productions of the two genders (e.g., Mulac et al., 2001; Mulac, 2006; Newman et al., 2008), showing that men were more likely to discuss money-

---

[1] We refer to COVID-19 by 'COVID' hereafter.

[2] `https://www.reddit.com/`

and occupation-related topics, focused on objects and impersonal matters, while women preferred discussion on family and social life, topics related to psychological and social processes. In the recent context, Thelwall and Thelwall (2020) found these observations persisted in COVID-19 tweets, with a male focus on sports and politics, and female focus on family and caring. In the prompted texts of the RWWD, van der Vegt and Kleinberg (2020) also found the expected M vs. F topical differences, with men talking more about the international impact of the pandemic, as well as governmental policy, and women more commonly discussing social aspects – family, friends, and solidarity. Moreover, van der Vegt and Kleinberg (2020) further found differences between the elicited short (tweet-sized) and longer essays, revealing the impact of the goal and size of the text on such analyses. Again, an open question remains concerning the topical distinctions between M and F authors in spontaneous productions without artificial restrictions on length.

Here, we aim to address the above gaps in the literature, by performing a comprehensive analysis of the similarities and differences between male and female language collected from the Reddit discussion platform. Our main corpus is a large collection of spontaneous COVID-related utterances by (self-reported) M and F authors. Importantly, we also collect productions on a wide variety of topics by the same set of authors as a 'baseline' dataset. First, using a multidimensional affective framework from psychology (Bradley and Lang, 1994), we draw on a recently-released dataset of human affective ratings of words Mohammad (2018) to support the emotional assessment of male and female posts in our datasets. Through this approach, we corroborate existing assumptions on differences in the emotional aspects of linguistic productions of men and women in the COVID corpus. Moreover, our use of a baseline dataset enables us to further show that these distinctions are amplified in the emotionally-intensive setting of COVID discussions compared to productions on other topics. Second, we take a topic modeling approach to demonstrate detectable distinctions in the range of topics discussed by the two genders in our COVID corpus, reinforcing (to some extent) assumptions on gender-related topical preferences, in this natural discourse in an emotionally-charged context.[3]

[3]All data and code is available at `https://github.com/ellarabi/covid19-demography`.

2 Datasets

As noted above, our goal is to analyze emotions and topics in spontaneous utterances that are relatively unconstrained by length. To that end, our main dataset comprises a large collection of spontaneous, COVID-related English utterances by male and female authors from the Reddit discussion platforms. As of May 2020, Reddit had over 430M active users, 1.2M topical threads (subreddits), and over 70% of its user base coming from English-speaking countries. Subreddits often encourage their subscribers to specify a meta-property (called a 'flair', a textual tag), projecting a small glimpse about themselves (e.g., political association, country of origin, age), thereby customizing their presence within a subreddit.

We identified a set of subreddits, such as 'r/askmen' and 'r/askwomen', where authors commonly self-report their gender, and extracted a set of unique user-ids of authors who specified male or female gender as a flair.[4] This process yielded the user-ids for $10,421$ males and $5,630$ females (as self-reported). Using this extracted set of ids, we collected COVID-related submissions and comments[5] from across the Reddit discussion platform for a period of 15 weeks, from February 1st through June 1st. COVID-related posts were identified as those containing one or more of a set of predefined keywords: 'covid', 'covid-19', 'covid19', 'corona', 'coronavirus', 'the virus', 'pandemic'. This process resulted in over 70K male and 35K female posts spanning $7,583$ topical threads; the male subcorpus contains 5.3M tokens and the female subcorpus 2.8M tokens. Figure 1 presents the weekly amount of COVID-related posts in the combined corpus, showing a peak in early-mid March (weeks 5–6).

Aiming at a comparative analysis between virus-related and 'neutral' (baseline) linguistic productions by men and women, we collected an additional dataset comprising a randomly sampled 10K posts per week by the same set of authors, totalling 150K posts for each gender. The baseline dataset contains 6.8M tokens in the male subcorpus and 5.3M tokens in the female subcorpus.

We use our COVID and baseline datasets for analysis of emotional differences as well as topical preferences in spontaneous productions by male

[4]Although gender can be viewed as a continuum rather than binary, we limit this study to the two most prominent gender markers in our corpus: male and female.

[5]For convenience, we refer to both initial submissions and comments to submissions as 'posts' hereafter.

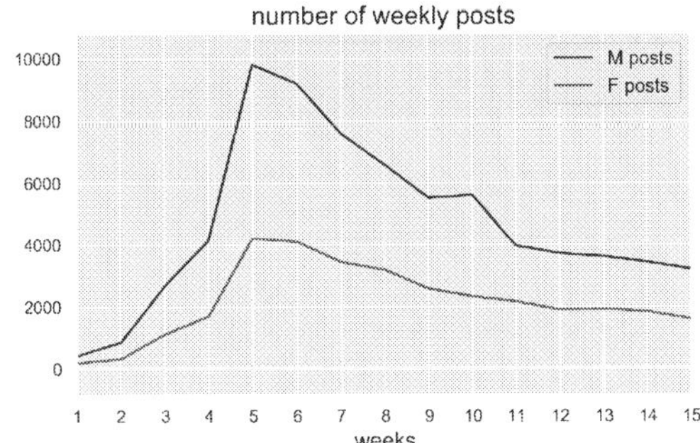

Figure 1: Weekly COVID-related posts by gender.

and female authors on Reddit. The ample size of the corpora facilitates analysis of distinctions in these two aspects between the two genders in their discourse on the pandemic, and as compared to non-COVID discussion.

3 Analysis of Emotional Dimensions

3.1 Methods

A common way to study emotions in psycholinguistics uses an approach that groups affective states into a few major dimensions, such as the Valence-Arousal-Dominance (VAD) affect representation, where *valence* refers to the degree of positiveness of the affect, *arousal* to the degree of its intensity, and *dominance* represents the level of control (Bradley and Lang, 1994). Computational studies applying this approach to emotion analysis have been relatively scarce due to the limited availability of a comprehensive resource of VAD rankings, with (to the best of our knowledge) no large-scale study on cross-gender language. Here we make use of the recently-released NRC-VAD Lexicon, a large dataset of human ratings of $20,000$ English words (Mohammad, 2018), in which each word is assigned V, A, and D values, each in the range [0–1]. For example, the word 'fabulous' is rated high on the valence dimension, while 'deceptive' is rated low. In this study we aim at estimating the VAD values of posts (typically comprising multiple sentences), rather than individual words; we do so by inferring the affective ratings of sentences using those of individual words, as follows.

Word embedding spaces have been shown to capture variability in emotional dimensions closely corresponding to valence, arousal, and dominance (Hollis and Westbury, 2016), implying that such semantic representations carry over information useful for the task of emotional affect assessment. Therefore, we exploit affective dimension ratings

assigned to individual words for supervision in extracting ratings of sentences. We use the model introduced by Reimers and Gurevych (2019) for producing word- and sentence-embeddings using Siamese BERT-Networks,[6] thereby obtaining semantic representations for the $20,000$ words in Mohammad (2018) as well as for sentences in our datasets. This model performs significantly better than alternatives (such as averaging over a sentence's individual word embeddings and using BERT encoding (Reimers and Gurevych, 2019)) on the SentEval toolkit, a popular evaluation toolkit for sentence embeddings (Conneau and Kiela, 2018).

Next, we trained beta regression models[7] (Zeileis et al., 2010) to predict VAD scores (dependent variables) of words from their embeddings (independent predictors), yielding Pearson's correlations of 0.85, 0.78, and 0.81 on a 1000-word held-out set for V, A, and D, respectively. The trained models were then used to infer VAD values for each sentence within a post using the sentence embeddings.[8] A post's final score was computed as the average of the predicted scores for each of its constituent sentences. As an example, the post *'most countries handled the covid-19 situation appropriately'* was assigned a low arousal score of 0.274, whereas a high arousal score of 0.882 was assigned to *'gonna shoot the virus to death!'*.

3.2 Results and Discussion

We compared V, A, and D scores of male posts to those of female posts, in each of the COVID and baseline datasets, using Wilcoxon rank sum tests. All differences were significant, and Cohen's d (Cohen, 2013) was used to find the effect size of these differences; see Table 1. We also compared the scores for each gender in the COVID dataset to their respective scores in the baseline dataset (discussed below). We further show, in Figure 2, the diachronic trends in VAD for M and F authors in the two sub-corpora: COVID and baseline.

First, Table 1 shows considerable differences between M and F authors in the baseline dataset for all three emotional dimensions (albeit a tiny effect size in valence), in line with established assumptions in this field (Burriss et al., 2007; Hoffman, 2008; Thelwall et al., 2010): women score higher in use of pos-

---

[6]We used the `bert-large-nli-mean-tokens` model, obtaining highest scores on a the STS benchmark.

[7]An alternative to linear regression in cases where the dependent variable is a proportion (in 0–1 range).

[8]We excluded sentences shorter than 5 tokens.

	COVID-related posts					Baseline posts				
	mean(M)	std(M)	mean(F)	std(F)	eff. size	mean(M)	std(M)	mean(F)	std(F)	eff. size
V	0.375	0.12	**0.388**	0.11	-0.120	0.453	0.14	**0.459**	0.14	-0.043
A	**0.579**	0.09	0.567	0.08	0.144	**0.570**	0.10	0.559	0.09	0.109
D	**0.490**	0.08	0.476	0.07	0.183	**0.486**	0.09	0.469	0.09	0.185

Table 1: Means of M and F posts for each affective dimension, and effect size of differences within each corpus. All differences significant at p<0.001. Highest mean score for each of V, A, D, in COVID and baseline, is boldfaced.

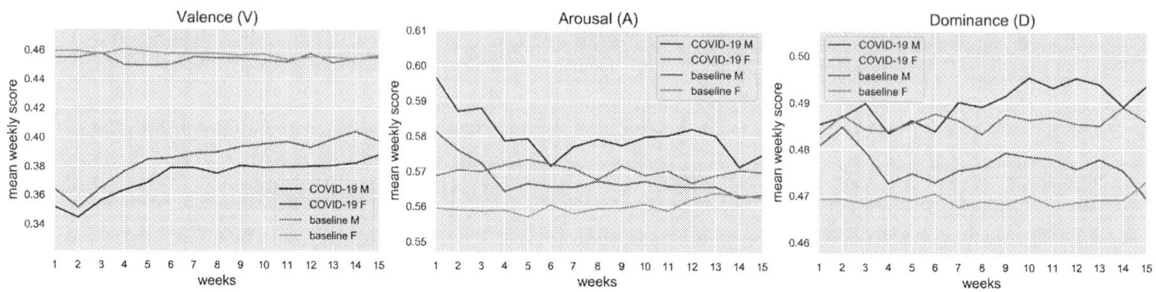

Figure 2: Diachronic analysis of valence (left), arousal (middle), and dominance (right) scores for Reddit data.

itive language, while men score higher on arousal and dominance. Interestingly, the cross-gender differences in V and A are amplified between baseline and COVID data, with an increase in effect size from 0.043 to 0.120 for V and 0.109 to 0.144 for A. By comparison, virtually no difference was detected in D between M and F authors in baseline vs. virus-related discussions. Thus we find that men seem to use more negative and emotionally-charged language when discussing COVID than women do – and to a greater degree than in non-COVID discussion – presumably indicating a grimmer outlook towards the pandemic. This finding is particularly interesting, given that van der Vegt and Kleinberg (2020) find that women self-report more negative emotion in reaction to the pandemic, and underlies the importance of analysis of implicit indications of affective state in spontaneous text.

COVID-related data trends (Figure 2) show comparatively low scores for valence and high scores for arousal in the early weeks of our analysis (February to mid-March). We attribute these findings to an increased level of alarm and uncertainty about the pandemic in its early stages, which gradually attenuated as the population learned more about the virus. As expected, both genders exhibit lower V scores in COVID discussions compared to baseline: Cohen's d effect size of -0.617 for M and -0.554 for F authors. Smaller, yet considerable, differences between the two sub-corpora exist also for A and D (0.095 and 0.047 for M, and 0.083 and 0.085, for F). These affective divergences from baseline show how emotionally-intensive is COVID-related discourse.

4   Analysis of Topical Distinctions

We study topical distinctions in male vs. female COVID-related discussions with two complementary analyses: (1) comparison of topics found by topic modelling over each of the M and F subcorpora separately, and (2) comparison of the distribution of dominant topics in M vs. F posts as derived from a topic model over the entire M+F dataset.

For each analysis, we used a publicly-available topic modeling tool (MALLET, McCallum, 2002). Each topic is represented by a probability distribution over the entire vocabulary, where terms more characteristic of a topic are assigned a higher probability.[9] A common way to evaluate a topic learned from a set of documents is by computing its *coherence score* – a measure reflecting its overall quality (Newman et al., 2010). We assess the quality of a learned model by averaging the scores of its individual topics – the *model* coherence score.

Analysis of Cross-gender Topics. Here we explore topical aspects of the productions of the two genders by comparing two topic models: one created using M posts, and another using F posts, in the COVID dataset. We selected the optimal number of topics for each set of posts by maximizing its model coherence score, resulting in 8 topics

---

[9]Prior to topic modeling we applied a preprocessing step including lemmatization of a post's text and filtering out stopwords (the 300 most frequent words in the corpus).

M-1	**M-2**	**M-3**	**M-4**	**F-1**	**F-2**	**F-3**	**F-4**
money	week	case	fuck	virus	feel	mask	week
economy	health	rate	mask	make	thing	hand	test
business	close	spread	claim	good	good	wear	hospital
market	food	hospital	news	thing	friend	woman	sick
crisis	open	week	post	vaccine	talk	food	patient
make	travel	month	comment	point	make	face	symptom
economic	supply	testing	call	happen	love	call	doctor
pandemic	store	social	article	human	parent	store	positive
lose	stay	lockdown	chinese	body	anxiety	close	start
vote	plan	measure	medium	study	read	stay	care

Table 2: Most coherent topics identified in male (**M-1**–**M-4**) and female (**F-1**–**F-4**) COVID-related posts.

	Topic	Keywords	Male	Female
1	**Economy**	money, business, make, month, food, economy, market, supply, store, cost	**0.17**	**0.10**
2	**Social**	feel, thing, live, good, make, friend, talk, love, hard, start	**0.07**	**0.26**
3	Distancing	close, social, health, open, plan, stay, travel, week, continue, risk	0.09	0.11
4	Virus	virus, kill, human, disease, study, body, spread, effect, similar, immune	0.11	0.07
5	Health (1)	mask, hand, stop, make, call, good, wear, face, person, woman	0.07	0.08
6	Health (2)	case, test, hospital, rate, spread, patient, risk, care, sick, testing	0.17	0.14
7	**Politics**	problem, issue, change, response, vote, policy, support, power, action, agree	**0.17**	**0.07**
8	Media	point, make, question, post, news, read, fact, information, understand, article	0.08	0.07
9	Misc.	good, start, thing, make, hour, stuff, play, pretty, find, easy	0.08	0.10

Table 3: Distribution of dominant topics in the COVID corpus. Entries in columns M(ale) and F(emale) represent the ratio of posts with the topic in that row as their main topic. Ratios are calculated for M and F posts separately (each of columns M and F sum to 1). Bolded topics indicate those with substantial differences between M and F.

for male and 7 topics for female posts (coherence scores of 0.48 and 0.46).

We examined the similarities and the differences across the two topical distributions by extracting the top 4 topics – those with the highest individual coherence scores – in each of the M and F models. Table 2 presents the 10 words with highest likelihood for these topics in each model; topics within each are ordered by decreasing coherence score (left to right). We can see that both genders are occupied with health-related issues (topics **M-3**, **F-1**, **F-4**), and the implications on consumption habits (topics **M-2**, **F-3**). However, clear distinctions in topical preference are also revealed by our analysis: men discuss economy/market and media-related topics (**M-1**, **M-4**), while women focus more on family and social aspects (**F-2**). Collectively these results show that the established postulates regarding gender-linked topical preferences are evident in spontaneous COVID-related discourse on Reddit.

Analysis of Dominance of Topics across Genders. We next performed a complementary analysis, creating a topic model over the combined male and female sub-corpora, yielding 9 topics.[10] We

calculate, for the two sets of M and F posts, the distribution of dominant topics – that is, for each of topics 1–9, what proportion of M (respectively F) posts had that topic as its first-ranked topic.

Table 3 reports the results; e.g., row 1 shows that the economy is the main topic of 17% of male posts, but only 10% of female posts. We see that males tend to focus more on economic and political topics than females (rows 1 and 7); conversely, females focus far more on social topics than did males (row 2). Once again, these findings highlight cross-gender topical distinctions in COVID discussions on Reddit in support of prior results.

5 Conclusions

A large body of studies spanning a range of disciplines has suggested (and corroborated) assumptions regarding the differences in linguistic productions of male and female speakers. Using a large dataset of COVID-related utterances by men and women on the Reddit discussion platforms, we show clear distinctions along emotional dimensions between the two genders, and demonstrate that these differences are amplified in emotionally-intensive discourse on the pandemic. Our analysis of topic modeling further highlights distinctions in topical preferences between men and women.

---

[10]We used the model with the 2nd-best number of topics (9, coherence score 0.432) as inspection revealed it to be more descriptive than the optimal number of topics (2, score 0.450).

Acknowledgments

This research was supported by NSERC grant RGPIN-2017-06506 to Suzanne Stevenson, and by an NSERC USRA to Jai Aggarwal.

References

Alaa Abd-Alrazaq, Dari Alhuwail, Mowafa Househ, Mounir Hamdi, and Zubair Shah. 2020. Top Concerns of Tweeters During the COVID-19 Pandemic: Infoveillance Study. *Journal of Medical Internet Research*, 22(4):e19016.

Margaret M Bradley and Peter J Lang. 1994. Measuring emotion: the self-assessment manikin and the semantic differential. *Journal of behavior therapy and experimental psychiatry*, 25(1):49–59.

Louisa Burriss, DA Powell, and Jeffrey White. 2007. Psychophysiological and subjective indices of emotion as a function of age and gender. *Cognition and emotion*, 21(1):182–210.

Jacob Cohen. 2013. *Statistical power analysis for the behavioral sciences*. Academic press.

Alexis Conneau and Douwe Kiela. 2018. SentEval: An Evaluation Toolkit for Universal Sentence Representations. *LREC 2018 - 11th International Conference on Language Resources and Evaluation*, pages 1699–1704.

Martin L Hoffman. 2008. Empathy and prosocial behavior. *Handbook of emotions*, 3:440–455.

Geoff Hollis and Chris Westbury. 2016. The principals of meaning: Extracting semantic dimensions from co-occurrence models of semantics. *Psychonomic Bulletin and Review*, 23(6):1744–1756.

Bennett Kleinberg, Isabelle van der Vegt, and Maximilian Mozes. in press. Measuring Emotions in the COVID-19 Real World Worry Dataset. Association for Computational Linguistics.

Moshe Koppel, Shlomo Argamon, and Anat Rachel Shimoni. 2002. Automatically categorizing written texts by author gender. *Literary and linguistic computing*, 17(4):401–412.

William Labov. 1990. The intersection of sex and social class in the course of linguistic change. *Language variation and change*, 2(2):205–254.

Robin Lakoff. 1973. Language and woman's place. *Language in society*, 2(1):45–79.

Qian Liu, Zequan Zheng, Jiabin Zheng, Qiuyi Chen, Guan Liu, Sihan Chen, Bojia Chu, Hongyu Zhu, Babatunde Akinwunmi, Jian Huang, Casper J. P. Zhang, and Wai-Kit Ming. 2020. Health Communication Through News Media During the Early Stage of the COVID-19 Outbreak in China: Digital Topic Modeling Approach. *Journal of Medical Internet Research*, 22(4):e19118. Company: Journal of Medical Internet Research Distributor: Journal of Medical Internet Research Institution: Journal of Medical Internet Research Label: Journal of Medical Internet Research Publisher: JMIR Publications Inc., Toronto, Canada.

May Oo Lwin, Jiahui Lu, Anita Sheldenkar, Peter Johannes Schulz, Wonsun Shin, Raj Gupta, and Yinping Yang. 2020. Global Sentiments Surrounding the COVID-19 Pandemic on Twitter: Analysis of Twitter Trends. *JMIR Public Health and Surveillance*, 6(2).

Andrew Kachites McCallum. 2002. MALLET: A machine learning for language toolkit.

Saif Mohammad. 2018. Obtaining reliable human ratings of valence, arousal, and dominance for 20,000 english words. In *Proceedings of the 56th Annual Meeting of the Association for Computational Linguistics (Volume 1: Long Papers)*, pages 174–184.

Anthony Mulac. 2006. *The gender-linked language effect: Do language differences really make a difference?* Lawrence Erlbaum Associates Publishers.

Anthony Mulac, James J Bradac, and Pamela Gibbons. 2001. Empirical support for the gender-as-culture hypothesis: An intercultural analysis of male/female language differences. *Human Communication Research*, 27(1):121–152.

David Newman, Jey Han Lau, Karl Grieser, and Timothy Baldwin. 2010. Automatic evaluation of topic coherence. In *Human Language Technologies: The 2010 Annual Conference of the North American Chapter of the Association for Computational Linguistics*, pages 100–108. Association for Computational Linguistics.

Matthew L Newman, Carla J Groom, Lori D Handelman, and James W Pennebaker. 2008. Gender differences in language use: An analysis of 14,000 text samples. *Discourse Processes*, 45(3):211–236.

Gregory Park, David Bryce Yaden, H Andrew Schwartz, Margaret L Kern, Johannes C Eichstaedt, Michael Kosinski, David Stillwell, Lyle H Ungar, and Martin EP Seligman. 2016. Women are warmer but no less assertive than men: Gender and language on facebook. *PloS one*, 11(5).

Nils Reimers and Iryna Gurevych. 2019. Sentence-BERT: Sentence Embeddings using Siamese BERT-Networks. *EMNLP-IJCNLP 2019 - 2019 Conference on Empirical Methods in Natural Language Processing and 9th International Joint Conference on Natural Language Processing, Proceedings of the Conference*, pages 3982–3992.

Jonathan Schler, Moshe Koppel, Shlomo Argamon, and James W Pennebaker. 2006. Effects of age and gender on blogging. In *AAAI spring symposium: Computational approaches to analyzing weblogs*, volume 6, pages 199–205.

H Andrew Schwartz, Johannes C Eichstaedt, Margaret L Kern, Lukasz Dziurzynski, Stephanie M Ramones, Megha Agrawal, Achal Shah, Michal Kosinski, David Stillwell, Martin EP Seligman, et al. 2013. Personality, gender, and age in the language of social media: The open-vocabulary approach. *PloS one*, 8(9):e73791.

Massimo Stella, Valerio Restocchi, and Simon De Deyne. 2020. #lockdown: Network-Enhanced Emotional Profiling in the Time of COVID-19. *Big Data and Cognitive Computing*, 4(2).

Mike Thelwall and Saheeda Thelwall. 2020. Covid-19 tweeting in English: Gender differences. *El Profesional de la Información*, 29(3).

Mike Thelwall, David Wilkinson, and Sukhvinder Uppal. 2010. Data mining emotion in social network communication: Gender differences in MySpace. *Journal of the American Society for Information Science and Technology*, 61(1):190–199.

Isabelle van der Vegt and Bennett Kleinberg. 2020. Women worry about family, men about the economy: Gender differences in emotional responses to covid-19. In *Proceedings of the 12th International Conference on Social Informatics*.

Achim Zeileis, Francisco Cribari-Neto, Bettina Grün, and I Kos-midis. 2010. Beta regression in r. *Journal of statistical software*, 34(2):1–24.

Cross-language sentiment analysis of European Twitter messages during the COVID-19 pandemic

Anna Kruspe
German Aerospace Center (DLR)
Institute of Data Science
Jena, Germany
anna.kruspe@dlr.de

Matthias Häberle
Technical University of Munich (TUM)
Signal Processing in Earth Observation (SiPEO)
Munich, Germany
matthias.haeberle@tum.de

Iona Kuhn
German Aerospace Center (DLR)
Institute of Data Science
Jena, Germany
iona.kuhn@dlr.de

Xiao Xiang Zhu
German Aerospace Center (DLR)
Remote Sensing Technology Institute (IMF)
Oberpfaffenhofen, Germany
xiaoxiang.zhu@dlr.de

Abstract

Social media data can be a very salient source of information during crises. User-generated messages provide a window into people's minds during such times, allowing us insights about their moods and opinions. Due to the vast amounts of such messages, a large-scale analysis of population-wide developments becomes possible.

In this paper, we analyze Twitter messages (tweets) collected during the first months of the COVID-19 pandemic in Europe with regard to their sentiment. This is implemented with a neural network for sentiment analysis using multilingual sentence embeddings. We separate the results by country of origin, and correlate their temporal development with events in those countries. This allows us to study the effect of the situation on people's moods. We see, for example, that lockdown announcements correlate with a deterioration of mood in almost all surveyed countries, which recovers within a short time span.

1 Introduction

The COVID-19 pandemic has led to a worldwide situation with a large number of unknowns. Many heretofore unseen events occurred within a short time span, and governments have had to make quick decisions for containing the spread of the disease. Due to the extreme novelty of the situation, the outcomes of many of these events have not been studied well so far. This is true with regards to their medical effect, as well as the effect on people's perceptions and moods.

First studies about the effect the pandemic has on people's lives are being published at the moment (e.g. Betsch et al., 2020), mainly focusing on surveys and polls. Naturally, such studies are limited to relatively small numbers of participants and focus on specific regions (e.g. countries).

In contrast, social media provides a large amount of user-created messages reflective of those users' moods and opinions. The issue with this data source is the difficulty of analysis - social media messages are extremely noisy and idiosyncratic, and the amount of incoming data is much too large to analyze manually. We therefore need automatic methods to extract meaningful insights.

In this paper, we describe a data set collected from Twitter during the months of December 2019 through April 2020, and present an automatic method for determining the sentiments contained in these messages. We then calculate the development of these sentiments over time, segment the results by country, and correlate them with events that took place in each country during those five months.

2 Related work

Since the pandemic outbreak and lockdown measures, numerous studies have been published to investigate the impact of the corona pandemic on

Twitter.

Feng and Zhou (2020) analyzed tweets from the US on a state and county level. First, they could detect differences in temporal tweeting patterns and found that people tweeting more about COVID-19 during working hours as the pandemic progressed. Furthermore, they conducted a sentiment analysis over time including an event specific subtask reporting negative sentiment when the 1000th death was announced and positive when the lockdown measures were eased in the states.

Lyu et al. (2020) looked into US-tweets which contained the terms "Chinese-virus" or "Wuhan-virus" referring to the COVID-19 pandemic to perform a user characterization. They compared the results to users that did not make use of such controversial vocabulary. The findings suggest that there are noticeable differences in age group, geolocation, or followed politicians.

Chen et al. (2020) focused on sentiment analysis and topic modelling on COVID-19 tweets containing the term "Chinese-virus" (controversial) and contrasted them against tweets without such terms (non-controversial). Tweets containing "Chinese-virus" discussing more topics which are related to China whereas tweets without such words stressing how to defend the virus. The sentiment analysis revealed for both groups negative sentiment, yet with a slightly more positive and analytical tone for the non-controversial tweets. Furthermore, they accent more the future and what the group itself can do to fight the disease. In contrast, the controversial group aiming more on the past and concentrate on what others should do.

3   Data collection

For our study, we used the freely available Twitter API to collect the tweets from December 2019 to April 2020. The free API allows streaming of 1% of the total tweet amount. To cover the largest possible area, we used a bounding box which includes the entire world. From this data, we sub-sampled 4,683,226 geo-referenced tweets in 60 languages located in the Europe. To create the Europe sample, we downloaded a shapefile of the earth[1], then we filtered by country performing a point in polygon test using the Python package *Shapely*[2]. Figure 1 depicts the Europe Twitter activity in total numbers.

Most tweets come from the U.K. Tweets are not filtered by topic, i.e. many of them are going to be about other topics than COVID-19. This is by design. As we will describe later, we also apply a simple keyword filter to detect tweets that are probably COVID-19-related for further analysis.

4   Analysis method

We now describe how the automatic sentiment analysis was performed, and the considerations involved in this method.

4.1   Sentiment modeling

In order to analyze these large amounts of data, we focus on an automatic method for sentiment analysis. We train a neural network for sentiment analysis on tweets. The text input layer of the network is followed by a pre-trained word or sentence embedding. The resulting embedding vectors are fed into a 128-dimensional fully-connected ReLU layer with 50% dropout, followed by a regression output layer with sigmoid activation. Mean squared error is used as loss. The model is visualized in figure 2.

This network is trained on the *Sentiment140* dataset (Go et al., 2009). This dataset contains around 1.5 million tweets collected through keyword search, and then annotated automatically by detecting emoticons. Tweets are determined to have positive, neutral, or negative sentiment. We map these sentiments to the values 1.0, 0.5, and 0.0 for the regression. Sentiment for unseen tweets is then represented on a continuous scale at the output.

We test variants of the model using the following pre-trained word- and sentence-level embeddings:

- A skip-gram version of *word2vec* (Mikolov et al., 2013) trained on the English-language Wikipedia[3]

- A multilingual version of BERT (Devlin et al., 2018) trained on Wikipedia data[4]

- A multilingual version of BERT trained on 160 million tweets containing COVID-19 keywords[5] (Müller et al., 2020)

---

[1] https://www.naturalearthdata.com/downloads/10m-cultural-vectors/10m-admin-0-countries/

[2] https://pypi.org/project/Shapely/

[3] https://tfhub.dev/google/Wiki-words-250/2

[4] https://tfhub.dev/tensorflow/bert_multi_cased_L-12_H-768_A-12/2

[5] https://tfhub.dev/digitalepidemiologylab/covid-twitter-bert/1

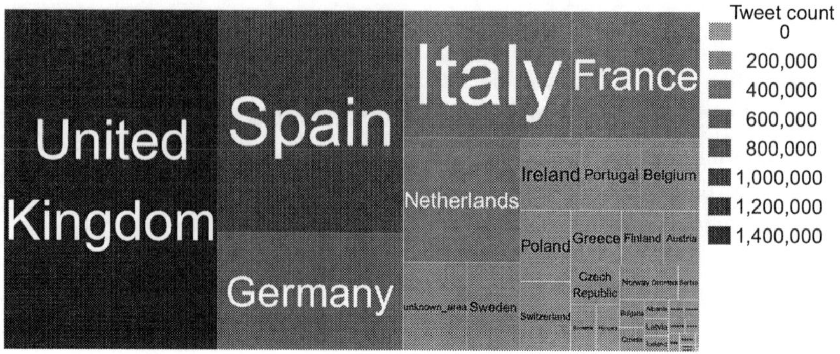

Figure 1: Treemap of Twitter activity in Europe during the time period of December 2019 to April 2020.

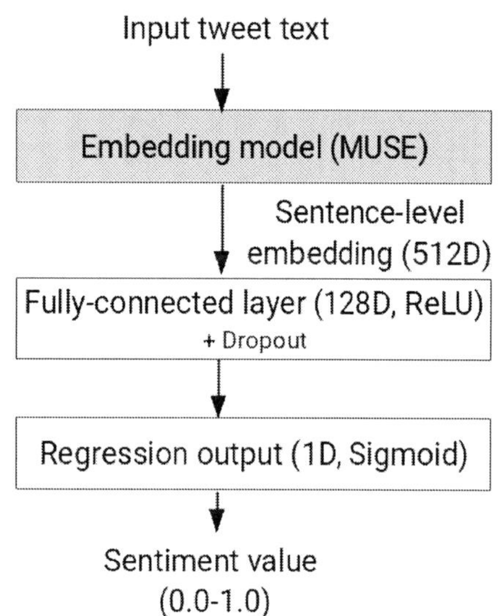

Figure 2: Architecture of the sentiment analysis model.

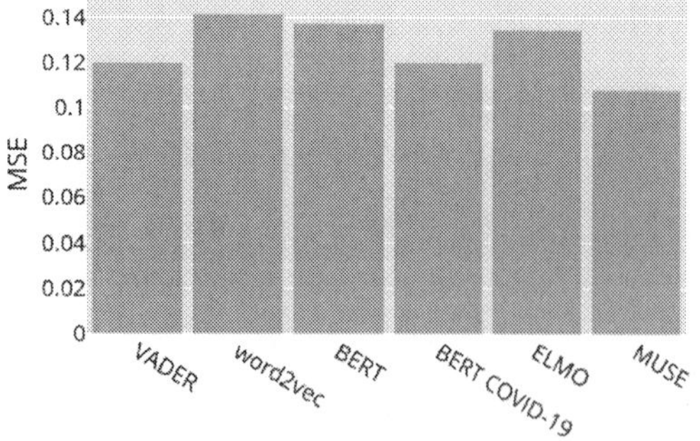

Figure 3: MSE for different models on the *Sentiment140* test dataset.

- An ELMO model (Peters et al., 2018) trained on the 1 Billion Word Benchmark dataset[6]

- The Multilingual Universal Sentence Encoder (MUSE)[7] (Yang et al., 2019)

We train each sentiment analysis model on the *Sentiment140* dataset for 10 epochs. Mean squared error results on the unseen test portion of the same dataset are shown in figure 3. For comparison, we also include an analysis conducted by VADER which is a rule-based sentiment reasoner designed for social media messages (Hutto and Gilbert, 2014).

Interestingly, most neural network results are in the range of the rule-based approach. BERT delivers better results than the *word2vec* model, with ELMO and the COVID-19-specific version also leading to improvements. However, the best result is achieved with the pre-trained multilingual USE model, which can embed whole sentences rather than (contextualized) words. We therefore perform the subsequent sentiment analysis with the MUSE-based model.

An interesting side note here is that the dataset only contains English-language tweets, but the sentence embedding is multilingual (for 16 languages). We freeze the embedding weights to prevent them from over-adapting to English. Due to the cross-lingual semantic representation capabilities of the pre-trained embedding, we expect the model to be able to detect sentiment in other languages just as well.

With the created model, we perform sentiment analysis on the 4.6 million tweets collected

---

[6]https://tfhub.dev/google/elmo/3
[7]https://tfhub.dev/google/universal-sentence-encoder-multilingual/3

from December to April, and then aggregate the results over time. This provides us with a representation of the development of Twitter messages' average sentiment over time. We specifically consider all collected tweets rather than just those determined to be topically related to COVID-19 because we are interested in the effect on people's moods in general, not just with regards to the pandemic. Additionally, we also filter the tweets by COVID-19-associated keywords, and analyze their sentiments as well. The chosen keywords are listed in figure 4.

4.2 Considerations

There are some assumptions implicit in this analysis method that we want to address here. First of all, we only consider tweets containing a geolocation. This applies to less than 1% of the whole tweet stream, but according to Sloan et al. (2013), the amount of geolocated tweets closely follows the geographic population distribution. According to Graham et al. (2014), there probably are factors determining which users share their locations and which ones do not, but there is no systematic study of these.

Other assumptions arise from the analysis method itself. For one, we assume that the model is able to extract meaningful sentiment values from the data. However, sentiment is subjective, and the model may be failing for certain constructs (e.g. negations, sarcasm). Additionally, modeling sentiment on a binary scale does not tell the whole story. "Positive" sentiment encompasses, for example, happy or hopeful tweets, "negative" angry or sad tweets, and "neutral" tweets can be news tweets, for example. A more finegrained analysis would be of interest in the future.

We also assume a somewhat similar perception of sentiment across languages. Finally, we assume that the detected sentiments as a whole are reflective of the mood within the community; on the other hand, mood is not quantifiable in the first place. All of these assumptions can be called into question. Nevertheless, while they may not be applicable for every single tweet, we hope to detect interesting effects on a large scale. When analyzing thousands of tweets within each time frame, random fluctuations become less likely. We believe that this analysis can provide useful insights into people's thoughts, and form an interesting basis for

future studies from psychological or sociological perspectives.

corona	コロナ	कोरोना
covid	冠狀病毒	โคโรน่า
wuhan	武漢	করোনা
koroona	ਕੋਰੋਨਾ	קורונה
корона	코로나	كورونا
κορων	կորոնա	ويروس
korona		

Figure 4: Keywords used for filtering the tweets (not case sensitive).

5 Results

In the following, we present the detected sentiment developments over time over-all and for select countries, and correlate them with events that took place within these months. Results for some other countries would have been interesting as well, but were not included because the main spoken language is not covered by MUSE (e.g. Sweden, Denmark). Others were excluded because there was not enough material available; we only analyze countries with at least 300,000 recorded tweets. As described in section 3, tweets are filtered geographically, not by language (i.e. Italian tweets may also be in other languages than Italian).

5.1 Over-all

In total, we analyzed around 4.6 million tweets, of which around 79,000 contained at least one COVID-19 keyword. Figure 5 shows the development of the sentiment over time for all tweets and for those with keywords, as well as the development of the number of keyworded tweets. The sentiment results are smoothed on a weekly basis (otherwise, we would be seeing a lot of movement during the week, e.g. an increase on the weekends). For the average over all tweets, we see a slight decrease in sentiment over time, indicating possibly that users' moods deteriorated over the last few months. There are some side effects that need to be considered here. For example, the curve rises slightly for holidays like Christmas and Easter (April 12). Interestingly, we see a clear dip around mid-March. Most European countries started implementing strong social distancing measures around this time. We will talk about this in more detail in the next sections.

We see that keywords were used very rarely before mid-January, and only saw a massive increase in usage around the beginning of March. Lately, usage has been decreasing again, indicating a loss of interest over time. Consequently, the sentiment analysis for keyword tweets is not expressive in the beginning. Starting with the more frequent usage in February, the associated sentiment drops massively, indicating that these tweets are now used in relation with the pandemic. Interestingly, the sentiment recovers with the increased use in March - it is possible that users were starting to think about the risks and handling of the situation in a more relaxed way over time. Still, the sentiment curve for keyword tweets lies significantly below the average one, which is to be expected for this all-around rather negative topic.

5.2 Analysis by country

We next aggregated the tweets by country as described in section 3 and performed the same analysis by country. The country-wise curves are shown jointly in figure 6. Comparing the absolute average sentiment values between countries is difficult as they may be influenced by the languages or cultural factors. However, the relative development is interesting. We see that all curves progress in a relatively similar fashion, with peaks around Christmas and Easter, a strong dip in the middle of March, and a general slow decrease in sentiment. In the following, we will have a closer look at each country's development. (Note that the keyword-only curves are cut of in the beginning for some countries due to a low number of keyword tweets).

5.2.1 Italy

Figure 7 shows the average sentiment for all Italian tweets and all Italian keyword tweets, as well as the development of keyword tweets in Italy. In total, around 400,000 Italian tweets are contained in the data set, of which around 12,000 have a keyword. Similar to the over-all curves described in section 5.1, the sentiment curve slowly decreases over time, keywords are not used frequently before the end of January, when the first cases in Italy were confirmed. Sentiment in the keyword tweets starts out very negative and then increases again. Interestingly, we see a dip in sentiment on March 9, which is exactly when the Italian lockdown was announced. Keywords were also used most frequently during that week. The dip is not visible in the keyword-only sentiment curve, suggesting that the negative sentiment was actually caused by the higher prevalence of coronavirus-related tweets.

5.2.2 Spain

For Spain, around 780,000 tweets were collected in total with around 14,000 keyword tweets. The curves are shown in figure 8. The heavier usage of keywords starts around the same time as in Italy, where the first domestic cases were publicized at the same time. The spike in keyword-only sentiment in mid-February is actually an artifact of the low number of keyworded tweets in combination with the fact that "corona" is a word with other meanings in Spanish (in contrast to the other languages). With more keyword mentions, the sentiment drops as in other countries.

From there onwards, the virus progressed somewhat slower in Spain, which is reflected in the curves as well. A lockdown was announced in Spain on March 14, corresponding to a dip in the sentiment curve. As with the Italian data, this dip is not present in the keyword-only sentiments.

5.2.3 France

Analyses for the data from France are shown in figure 9. For France, around 309,000 tweets and around 4,600 keyword tweets were collected. Due to the lower number of data points, the curves are somewhat less smooth. Despite the first European COVID-19 case being detected in France in January, cases did not increase significantly until the end of February, which once again is also seen in the start of increased keyword usage here. The French lockdown was announced on March 16 and extended on April 13, both reflected in dips in the sentiment curve. Towards the end of the considered period, keyword-only sentiment actually starts to increase, which is also seen in Italy and Germany. This could indicate a shift to a more hopeful outlook with regards to the pandemic.

5.2.4 Germany

For Germany, around 415,000 tweets and around 5,900 keyword tweets were collected. The analysis results are shown in figure 10. After very few first cases at the end of January, Germany's case count did not increase significantly until early March, which is again when keyword usage increased. The decrease in the sentiment curve actually arrives around the same time as in France and Spain, which is a little surprising because social distancing measures were not introduced by the government until March 22 (extended on March 29). German users were likely influenced by the situation in their neighboring countries here. In

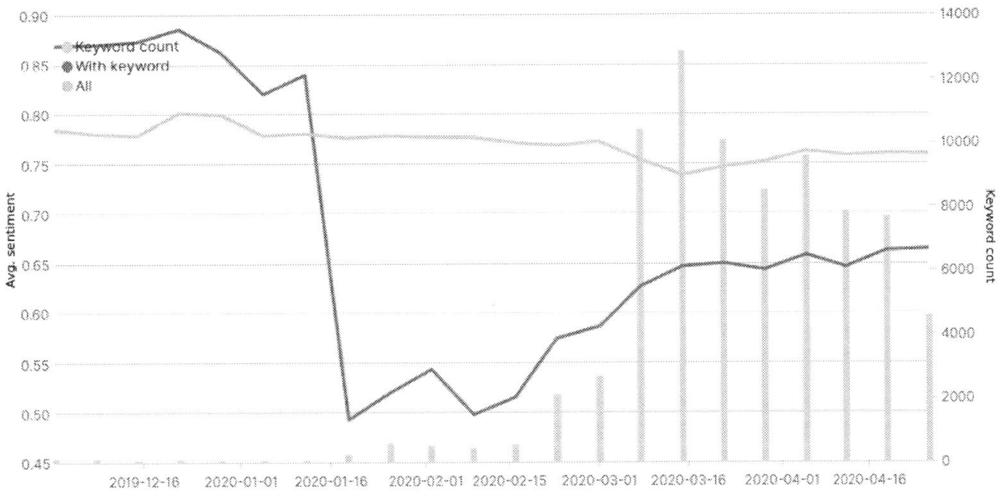

Figure 5: Development of average sentiment for all tweets and for tweets containing COVID-19 keywords, and development of number of tweets containing COVID-19 keywords.

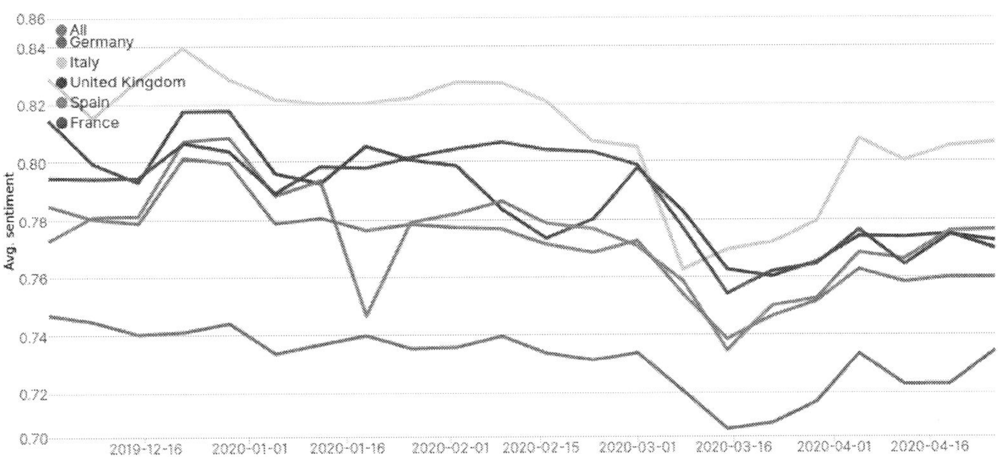

Figure 6: Development of average sentiment over time by country (all tweets).

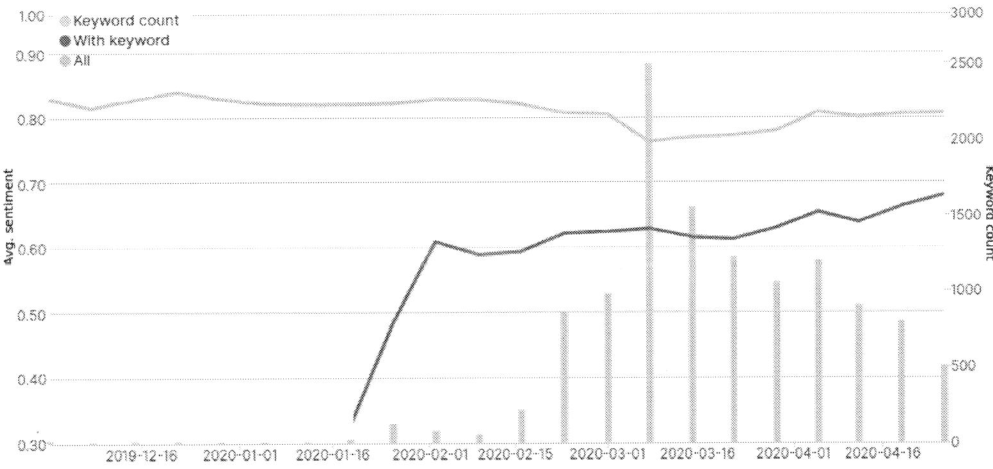

Figure 7: Italy: Development of average sentiment for all tweets and for tweets containing COVID-19 keywords, and development of number of tweets containing COVID-19 keywords.

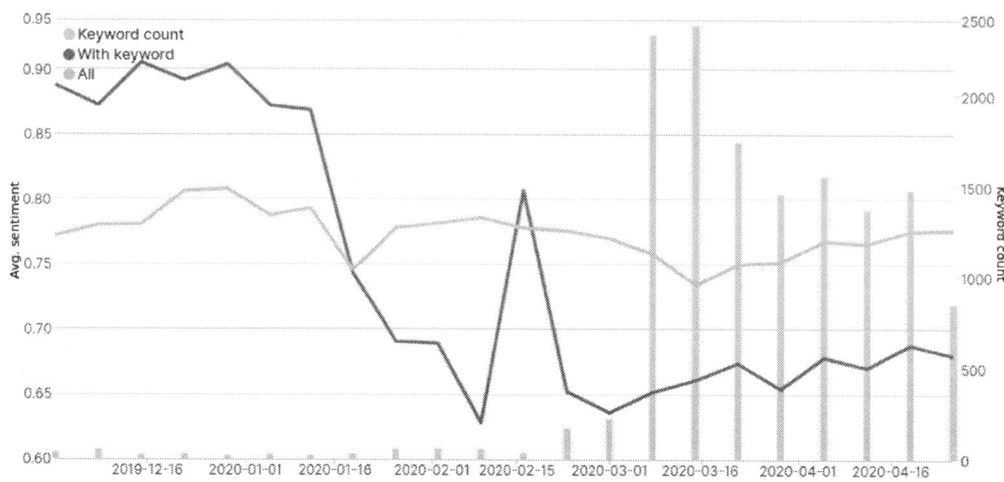

Figure 8: Spain: Development of average sentiment for all tweets and for tweets containing COVID-19 keywords, and development of number of tweets containing COVID-19 keywords.

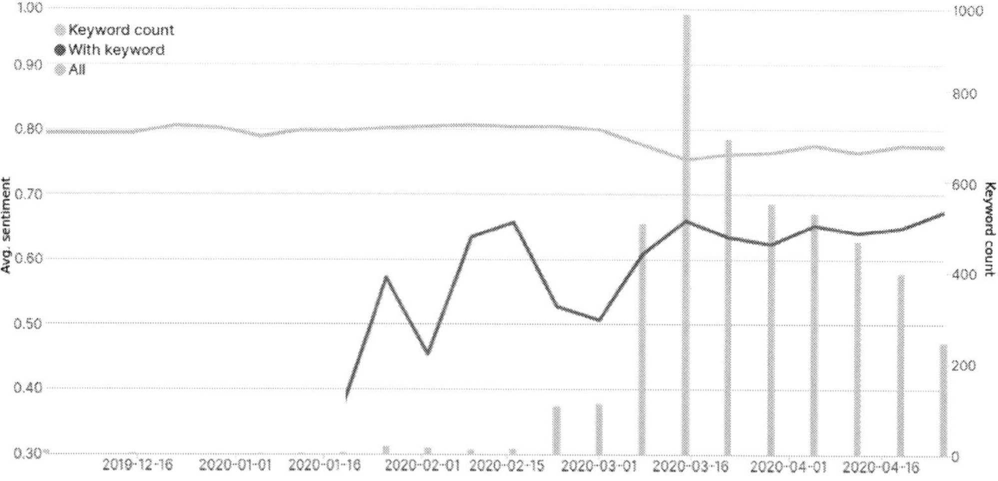

Figure 9: France: Development of average sentiment for all tweets and for tweets containing COVID-19 keywords, and development of number of tweets containing COVID-19 keywords.

general, the curve is flatter than in other countries. One possible reason for this might be the lower severity of measures in Germany, e.g. there were no strict curfews.

In contrast to all other considered countries, the keyword-only sentiment curve is not significantly below the sentiment curve for all tweets in Germany after the beginning of March. There are some possible explanations for this. For one, governmental response to the situation was generally applauded in Germany (Betsch et al., 2020), and, as mentioned above, was not as strict as in other countries, possibly not impacting people as much. On the other hand, the over-all German curve is lower than its counterparts from other countries, i.e. German tweets have lower average sentiment values in general, possibly caused by cultural factors.

5.2.5 United Kingdom

Curves for the United Kingdom are shown in figure 11, calculated on around 1,380,000 tweets including around 22,000 keyword tweets. Higher keyword usage starts somewhat earlier here than expected in February, whereas a significant increase in cases did not occur until March. Once again, keyword-only sentiment starts out very negative and then increases over time.

The British government handled the situation somewhat differently. In early March, only recommendations were given, and a lockdown was explicitly avoided to prevent economic consequences. This may be a cause for the sentiment peak seen at this time. However, the curve falls until mid-March, when other European countries did implement lockdowns. The government finally did announce a lockdown starting on March 26. This did not lead to a significant change in average sentiment anymore, but in contrast with other countries, the curve does not swing back to a significantly more positive level in the considered period, and actually decreases towards the end.

6 Conclusion

In this paper, we presented the results of a sentiment analysis of 4.6 million geotagged Twitter messages collected during the months of December 2019 through April 2020. This analysis was performed with a neural network trained on an unrelated Twitter sentiment data set. The tweets were then tagged with sentiment on a scale from 0 to 1 using this network. The results were aggregated by country, and averaged over time. Additionally, the sentiments of tweets containing COVID-19-related keywords were aggregated separately.

We find several interesting results in the data. First of all, there is a general downward trend in sentiment in the last few months corresponding to the COVID-19 pandemic, with clear dips at times of lockdown announcements and a slow recovery in the following weeks in most countries. COVID-19 keywords were used rarely before February, and correlate with a rise in cases in each country. The sentiment of keyworded tweets starts out very negative at the beginning of increased keyword usage, and becomes more positive over time. However, it remains significantly below the average sentiment in all countries except Germany. Interestingly, there is a slight upward development in sentiment in most countries towards the end of the considered period.

7 Future work

We will continue this study by also analyzing the development in the weeks since May 1st and the coming months. More countries will also be added. It will be very interesting to compare the shown European results to those of countries like China, South Korea, Japan, New Zealand, or even individual US states, which were impacted by the pandemic at different times and in different ways, and where the governmental and societal response was different from that of Europe.

There are also many other interesting research questions that could be answered on a large scale with this data - for example, regarding people's trust in published COVID-19 information, their concrete opinions on containment measures, or their situation during an infection. Other data sets have also been published in the meantime, including ones that contains hundreds of millions of tweets at the time of writing (e.g. Qazi et al., 2020; Banda et al., 2020). These data sets are much larger because collection was not restricted to geotagged tweets In Qazi et al. (2020), geolocations were instead completed from outside sources.

These studies could also be extended to elucidate more detailed factors in each country. One possibility here is an analysis of Twitter usage and tweet content by country. Another, as mentioned above, lies in moving from the binary sentiment scale to a more complex model.

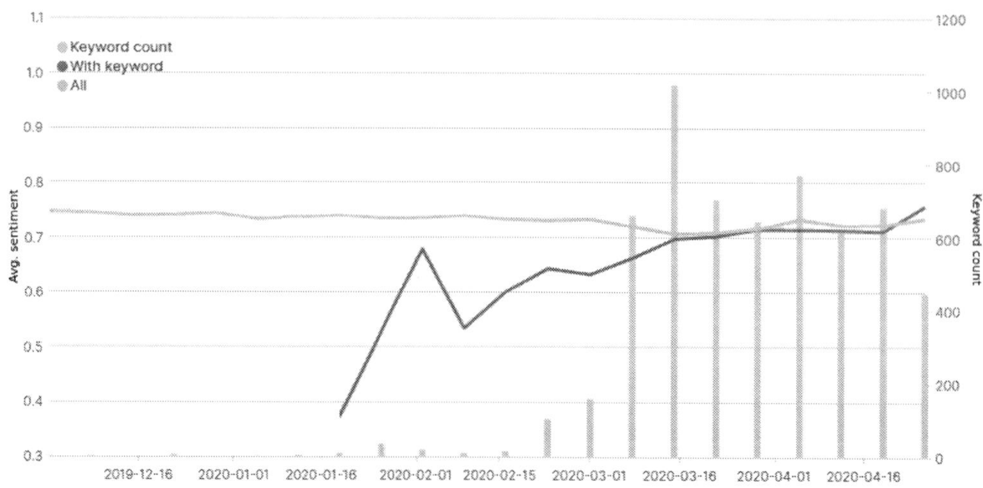

Figure 10: Germany: Development of average sentiment for all tweets and for tweets containing COVID-19 keywords, and development of number of tweets containing COVID-19 keywords.

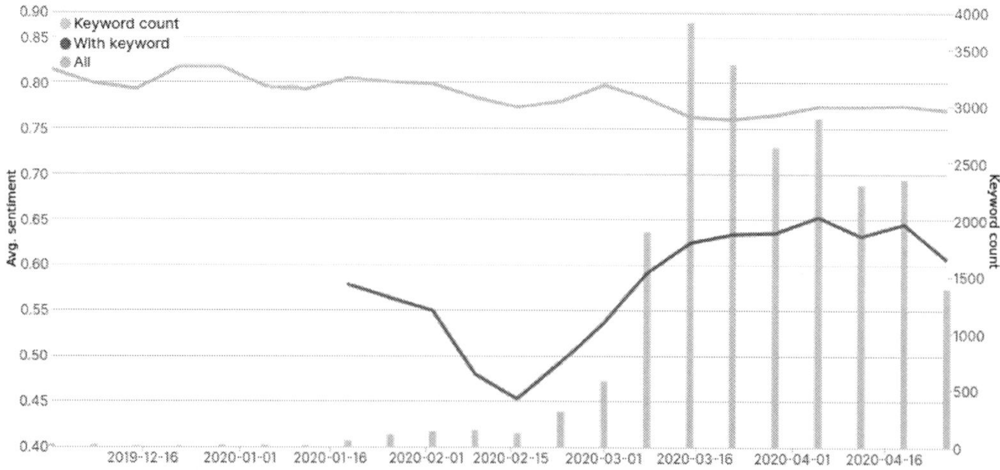

Figure 11: United Kingdom: Development of average sentiment for all tweets and for tweets containing COVID-19 keywords, and development of number of tweets containing COVID-19 keywords.

References

Juan M. Banda, Ramya Tekumalla, Guanyu Wang, Jingyuan Yu, Tuo Liu, Yuning Ding, Katya Artemova, Elena Tutubalina, and Gerardo Chowell. 2020. A large-scale COVID-19 Twitter chatter dataset for open scientific research - an international collaboration. https://doi.org/10.5281/zenodo.3723939.

Cornelia Betsch, Lars Korn, Lisa Felgendreff, Sarah Eitze, Philipp Schmid, Philipp Sprengholz, Lothar Wieler, Patrick Schmich, Volker Stollorz, Michael Ramharter, Michael Bosnjak, Saad B. Omer, Heidrun Thaiss, Freia De Bock, Ursula von Rüden, Cara Ebert, Janina Steinert, and Martin Bruder. 2020. German COVID-19 Snapshot MOnitoring (COSMO Germany). https://www.psycharchives.org/handle/20.500.12034/2398.

Long Chen, Hanjia Lyu, Tongyu Yang, Yu Wang, and Jiebo Luo. 2020. In the eyes of the beholder: Sentiment and topic analyses on social media use of neutral and controversial terms for covid-19. arXiv:2004.10225.

Jacob Devlin, Ming-Wei Chang, Kenton Lee, and Kristina Toutanova. 2018. BERT: pre-training of deep bidirectional transformers for language understanding. *CoRR*, abs/1810.04805.

Yunhe Feng and Wenjun Zhou. 2020. Is Working From Home The New Norm? An Observational Study Based on a Large Geo-tagged COVID-19 Twitter Dataset. arXiv:2006.08581.

Alec Go, Richa Bhayani, and Lei Huang. 2009. Twitter sentiment classification using distant supervision. Technical report, Stanford University.

Mark Graham, A. Scott Hale, and Devin Gaffney. 2014. Where in the World are You? Geolocation and Language Identification in Twitter. *The Professional Geographer*.

C.J. Hutto and Eric Gilbert. 2014. VADER: A parsimonious rule-based model for sentiment analysis of social media text. In *Eighth International Conference on Weblogs and Social Media (ICWSM-14)*.

Hanjia Lyu, Long Chen, Yu Wang, and Jiebo Luo. 2020. Sense and Sensibility: Characterizing Social Media Users Regarding the Use of Controversial Terms for COVID-19. arXiv:2004.06307.

Tomas Mikolov, Kai Chen, Greg Corrado, and Jeffrey Dean. 2013. Efficient estimation of word representations in vector space. In *International Conference on Learning Representations (ICLR)*, Scottsdale, AZ, USA.

Martin Müller, Marcel Salathé, and Per E. Kummervold. 2020. COVID-Twitter-BERT: A Natural Language Processing Model to Analyse COVID-19 Content on Twitter. arXiv:2005.07503.

Matthew E. Peters, Mark Neumann, Mohit Iyyer, Matt Gardner, Christopher Clark, Kenton Lee, and Luke Zettlemoyer. 2018. Deep contextualized word representations. *CoRR*, abs/1802.05365.

Umair Qazi, Muhammad Imran, and Ferda Ofli. 2020. Geocov19: A dataset of hundreds of millions of multilingual covid-19 tweets with location information. *ACM SIGSPATIAL Special*, 12(1):6–15.

Luke Sloan, Jeffrey Morgan, William Housley, Matthew Williams, Adam Edwards, Pete Burnap, and Omer Rana. 2013. Knowing the tweeters: Deriving sociologically relevant demographics from twitter. *Sociological Research Online*, 18(3):7.

Yinfei Yang, Daniel Cer, Amin Ahmad, Mandy Guo, Jax Law, Noah Constant, Gustavo Hernandez Abrego, Steve Yuan, Chris Tar, Yun-Hsuan Sung, Brian Strope, and Ray Kurzweil. 2019. Multilingual universal sentence encoder for semantic retrieval. arXiv:1907.04307.

Cross-lingual Transfer Learning for COVID-19 Outbreak Alignment

Sharon Levy and **William Yang Wang**
University of California, Santa Barbara
Santa Barbara, CA 93106
{sharonlevy,william}@cs.ucsb.edu

Abstract

The spread of COVID-19 has become a significant and troubling aspect of society in 2020. With millions of cases reported across countries, new outbreaks have occurred and followed patterns of previously affected areas. Many disease detection models do not incorporate the wealth of social media data that can be utilized for modeling and predicting its spread. It is useful to ask, can we utilize this knowledge in one country to model the outbreak in another? To answer this, we propose the task of cross-lingual transfer learning for epidemiological alignment. Utilizing both macro and micro text features, we train on Italy's early COVID-19 outbreak through Twitter and transfer to several other countries. Our experiments show strong results with up to 0.85 Spearman correlation in cross-country predictions.

1 Introduction

During the COVID-19 pandemic, society was brought to a standstill, affecting many aspects of our daily lives. With increased travel due to globalization, it is intuitive that countries have followed earlier affected regions in outbreaks and measures to contain to them (Cuffe and Jeavans, 2020).

A unique form of information that can be used for modeling disease propagation comes from social media. This can provide researchers with access to unfiltered data with clues as to how the pandemic evolves. Current research on the COVID-19 outbreak concerning social media includes word frequency and sentiment analysis of tweets (Rajput et al., 2020) and studies on the spread of misinformation (Kouzy et al., 2020; Singh et al., 2020). Social media has also been utilized for other disease predictions. Several papers propose models to identify tweets in which the author or nearby person has the attributed disease (Kanouchi et al., 2015; Aramaki et al., 2011; Lamb et al., 2013; Kitagawa et al., 2015). Iso et al. (2016) and Huang et al.

(2016) utilize word frequencies to align tweets to disease rates. A shortcoming of the above models is they do not consider how one region's outbreak may relate to another. Many of the proposed models also rely on lengthy keyword lists or syntactic features that may not generalize across languages. Text embeddings from models such as multilingual BERT (mBERT) (Devlin et al., 2019) and LASER (Artetxe and Schwenk, 2019) can allow us to combine features and make connections across languages for semantic alignment.

We present an analysis of Twitter usage for cross-lingual COVID-19 outbreak alignment. We study the ability to correlate social media tweets across languages and countries in a pandemic scenario. Based on this demonstration, researchers can study various cross-cultural reactions to the pandemic on social media. We aim to analyze how one country's tweets align with its own outbreak and if those same tweets can be used to predict the state of another country. This can allow us to determine how actions taken to contain the outbreak can transfer across countries with similar measures. We show that we can achieve strong results with cross-lingual transfer learning.

Our contributions include:

- We formulate the task of cross-lingual transfer learning for epidemiological outbreak alignment across countries.

- We are the first to investigate state-of-the-art cross-lingual sentence embeddings for cross-country epidemiological outbreak alignment. We propose joint macro and micro reading for multilingual prediction.

- We obtain strong correlations in domestic and cross-country predictions, providing us with evidence that social media patterns in relation to COVID-19 transcend countries.

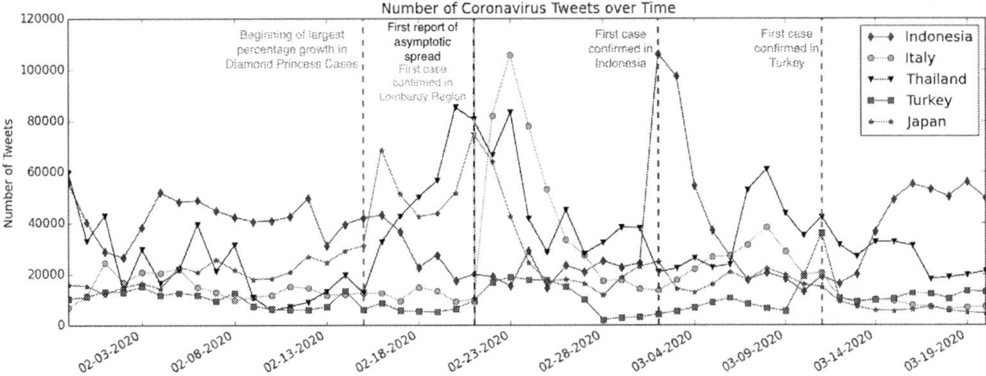

Figure 1: Timeline of COVID-19-related tweets, from COVID-19 dataset (Chen et al., 2020), in various languages. The peaks are marked by events relating to each language's main country's initial outbreak.

2 Twitter and COVID-19

2.1 Problem Formulation

An intriguing question in the scope of epidemiological research is: can atypical data such as social media help us model an outbreak? To study this, we utilize Twitter as our source, since users primarily post textual data and in real-time. Furthermore, Twitter users transcend several countries, which is beneficial as COVID-19 is analyzed by researchers and policymakers on a country by country basis (Kaplan et al., 2020). Our motivation in this paper is the intuition that social media users can provide us with indicators of an outbreak during the COVID-19 pandemic. In this case, we reformulate our original question: can we align Twitter with a country's COVID-19 outbreak and apply the learned information to other countries?

2.2 Data

We utilize the COVID-19 Twitter dataset (Chen et al., 2020), comprised of millions of tweets in several languages. These were collected through Twitter's streaming API and Tweepy[1] by filtering for 22 specific keywords and hashtags related to COVID-19 such as Coronavirus, Wuhanlockdown, stayathome, and Pandemic. We consider tweets starting from February 1st, 2020 to April 30th, 2020, and filter for tweets written in Italian, Indonesian, Turkish, Japanese, and Thai. Specifically, we filter for languages that are primarily spoken in only one country, as opposed to languages such as English and Spanish that are spoken in several countries. In Table 1, we show dataset statistics describing total tweet counts for each country along

---

[1]https://www.tweepy.org/

	Italy	Thailand	Japan	Turkey	Indonesia
Pre	1.3M	2.2M	2.2M	960K	3.2M
Post	103K	6.9K	61K	96K	309K

Table 1: Dataset statistics in each country before (Pre) and after (Post) the tweet filter process described in Section 2.5.

with counts after our filtering process described later in Section 2.5. When aligning tweets with each country's outbreak, we utilize the COVID-19 Dashboard by the CSSE at Johns Hopkins University (Dong et al., 2020) for daily confirmed cases from each country. Since the COVID-19 pandemic is still in its early stages at the time of writing this paper, sample sizes are limited. Therefore, our experiments have the following time cut settings: train in February and March and test in April (I), train in February and test in March and April (II), train in February and test in March (III), and train in March and test in April (IV).

2.3 Can Twitter detect the start of a country's outbreak?

We start by investigating a basic feature in our dataset: tweet frequency. We plot each country's tweet frequency in Figure 1. There is a distinct peak within each country, corresponding to events within each country signaling initial outbreaks, denoted by the vertical lines. These correlations indicate that even a standard characteristic such as tweet frequency can align with each country's outbreak and occurs across several countries. Given this result, we further explore other tweet features for epidemiological alignment.

2.4 Cross-Lingual Transfer Learning

We determine that it is most helpful for researchers to first study regions with earlier outbreaks to make assumptions on later occurrences in other locations. In this case, Italy has the earliest peak in cases. When aligning outbreaks from two different countries, we experiment with the transfer learning setting. We train on Italy's data and test on the remaining countries. We attempt to answer whether we can build a model that correlates the day's tweets with the number of cases in a given country and if we can apply this trained model to tweets and cases in a new country with a different language and culture.

We present this as a regression problem in which we map our input text features $\mathbf{x} \in \mathbb{R}^n$ to the output $\mathbf{y} \in \mathbb{R}$. Our ground-truth output $\mathbf{y}$ is presented in two scenarios in our experiments: total cases and daily new cases. The former considers all past and current reported cases while the latter consists of only cases reported on a specific day. The predicted output $\hat{\mathbf{y}}$ is compared against ground truth $\mathbf{y}$. During training and test time, we utilize support vector regression for our model and concatenate the chosen features as input each day. Due to different testing resources, criteria, and procedures, there are some offsets in each countries' official numbers. Therefore, we follow related disease prediction work and evaluate predictions with Spearman's correlation (Hogg et al., 2005) to align our features with official reported cases.

2.5 Creating a Base Model

In the wake of the COVID-19 crisis, society has adopted a new vocabulary to discuss the pandemic (Katella, 2020). Quarantine and lockdown have become standard words in our daily conversations. Therefore, we ask: are there specific features that indicate the state of an outbreak?

Which features can we utilize for alignment?
We create a small COVID-19-related keyword list consisting of lockdown, quarantine, social distancing, epidemic, and outbreak and translate these words into Italian. We include the English word "lockdown" as it has been used in other countries' vocabularies. We aim to observe which, if any, of these words align with Italy's outbreak. In addition to word frequencies, we also utilize mBERT and LASER to extract tweet representations for semantic alignment. We remove duplicate tweets, retweets, tweets with hyperlinks, and tweets dis-

Cases	Embed	Time Setting			
		I	II	III	IV
Total	mBERT	**0.880**	**0.947**	**0.769**	**0.880**
	LASER	0.879	0.946	0.766	0.879
New	mBERT	**0.805**	0.416	0.718	0.794
	LASER	0.800	**0.490**	**0.723**	**0.800**

Table 2: Italy's Spearman correlation results with total and daily case count prediction for mBERT and LASER (Embed). Time settings are defined in 2.2. We bold the highest correlations within each case setting.

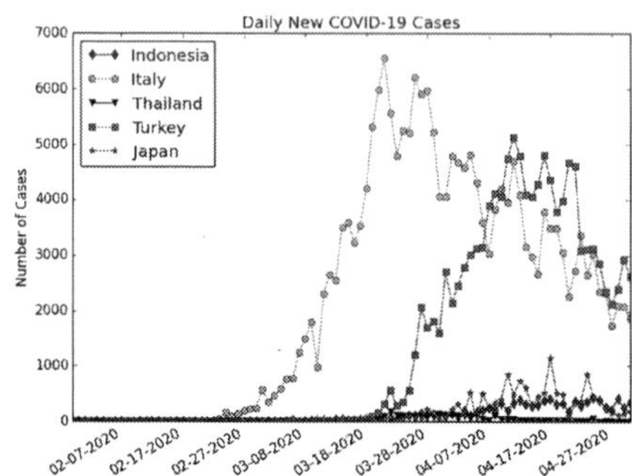

Figure 2: Distribution of new daily COVID-19 cases in Italy, Turkey, Thailand, Japan, and Indonesia. Daily case counts come from COVID-19 Dashboard by CSSE at Johns Hopkins University (Dong et al., 2020).

cussing countries other than Italy (tweets with other country names) in order to focus more on personal narratives within the country. Using the sentence encoding service bert-as-a-service (Xiao, 2018), we extract fixed-length representations for each tweet. We explore two options for our tweet representations: average-pooling and max-pooling. Our final feature consists of daily tweet frequency after filtering.

Can tweet text align with confirmed cases?
We combine combinations of our frequency features with our tweet embeddings and show results in Table 2. Through manual tuning, we find our strongest model (polynomial kernel) contained the English keyword lockdown and averaged tweet representations from mBERT for the total case scenario. When aligning to new cases, the best model (sigmoid kernel) contained the English keyword lockdown and max-pooled LASER embeddings. While mBERT and LASER provide very little difference in alignment to total cases, LASER is noticeably stronger in the new case setting, particularly in II. For the total case setting, our predictions show strong alignment with ground truth, which is monotonically increasing, in all time settings. When measuring new daily cases, the correlations

Setting	Thailand	Japan	Turkey	Indonesia
I	0.200	-.300	.188	-.316
II	0.696	0.543	0.715	0.285
III	0.823	0.856	0.679	0.925
IV	0.196	-.300	0.188	-.316
V	0.859	0.649	0.817	0.722

Table 3: Cross-lingual transfer learning Spearman correlation with total case counts while training with Italy data. Time settings are defined in 2.2.

Setting	Thailand	Japan	Turkey	Indonesia
I	-.022	0.130	-.368	0.416
II	0.277	0.273	0.426	0.332
III	0.661	0.262	0.255	0.407
IV	-.043	0.127	-.375	0.416
V	0.755	0.515	0.745	0.742

Table 4: Cross-lingual transfer learning Spearman correlation with new daily case counts while training with Italy data. Time settings are defined in 2.2.

are weaker in II. We find that Italy's new cases form a peak in late March, as shown in Figure 2. As a result, there is a distribution shift when training on February data only (tail of the distribution) and testing in March and April.

2.6 Cross-Lingual Prediction

While we can align historical data to future cases within Italy, researchers may not have enough data to train models for each country. Therefore we ask, can we use Italy's outbreak to predict the outbreak of another country? In particular, we determine whether users from two different countries follow similar patterns of tweeting during their respective pandemics and how well we can align the two. We follow the same tweet preprocessing methodology described in Section 2.5 and the timeline cuts for training and testing defined in Section 2.2. We also add another time setting (V): training in February, March, and April and testing all three months. This serves as an upper bound for our correlations, indicating how well the general feature trends align between the two countries and their outbreaks.

Can we transfer knowledge to other countries?
We show our results for the total and new daily case settings in Tables 3 and 4. All of the test countries have strong correlations in time setting V for both case settings. Since this is used as an upper bound, we can deduce that tweets across countries follow the same general trend in relation to reported cases. When examining the other time settings, it is clear that Italy transfers well in II and III for the total

case setting. As these train in February only, this shows us that transferring knowledge works better in times of more linear case increases, rather than during peaks, which becomes unstable. Times I through IV generally do not perform as well in the new case setting, though II and III primarily have higher correlations.

Why does Indonesia differ? It is noticeable that Indonesia aligns better with new daily cases in times I through IV, as opposed to the other countries. When examining Figure 2, we find that Indonesia is the only country that had not yet reached a peak in new daily cases by the end of April, and is steadily increasing. Meanwhile, the other countries follow normal distributions like Italy. However, given that we train our model on February and March data, it does not learn information on post-peak trends and cannot generalize well to these scenarios that occur in April in the other countries.

What can we learn from our results? Overall, transfer learning in the total case setting leads to stronger correlations with case counts. While results show that training in February and testing in March and/or April works best, our results for V's upper bound correlation show that weaker correlations can be due to the limited sample sizes we have from the start of the pandemic. Additionally, training in February, March, and April in Italy allows us to model a larger variety of scenarios during the pandemic, with samples during pre, mid, and post-peak. Therefore, as we obtain more data every day, we can build stronger models that can generalize better to varying distributions of cases and align outbreaks across countries that can fully reach their upper bound correlations and beyond.

3 Conclusion

In this paper, we performed an analysis of cross-lingual transfer learning with Twitter data for COVID-19 outbreak alignment using cross-lingual sentence embeddings and keyword frequencies. We showed that even with our limited sample sizes, we can utilize knowledge of countries with earlier outbreaks to correlate with cases in other countries. With larger sample sizes and when training on a variety of points during the outbreak, we can obtain stronger correlations to other countries. We hope our analysis can lead to future integration of social media in epidemiological prediction across countries, enhancing outbreak detection systems.

References

Eiji Aramaki, Sachiko Maskawa, and Mizuki Morita. 2011. Twitter catches the flu: Detecting influenza epidemics using twitter. In *Proceedings of the 2011 Conference on Empirical Methods in Natural Language Processing*, pages 1568–1576, Edinburgh, Scotland, UK. Association for Computational Linguistics.

Mikel Artetxe and Holger Schwenk. 2019. Massively multilingual sentence embeddings for zero-shot cross-lingual transfer and beyond. *Transactions of the Association for Computational Linguistics*, 7:597–610.

Emily Chen, Kristina Lerman, and Emilio Ferrara. 2020. Covid-19: The first public coronavirus twitter dataset. *arXiv preprint arXiv:2003.07372*.

Robert Cuffe and Christine Jeavans. 2020. How the uk's coronavirus epidemic compares to other countries. *BBC News*.

Jacob Devlin, Ming-Wei Chang, Kenton Lee, and Kristina Toutanova. 2019. BERT: Pre-training of deep bidirectional transformers for language understanding. In *Proceedings of the 2019 Conference of the North American Chapter of the Association for Computational Linguistics: Human Language Technologies, Volume 1 (Long and Short Papers)*, pages 4171–4186, Minneapolis, Minnesota. Association for Computational Linguistics.

Ensheng Dong, Hongru Du, and Lauren Gardner. 2020. An interactive web-based dashboard to track covid-19 in real time. *The Lancet infectious diseases*, 20(5):533–534.

Robert V Hogg, Joseph McKean, and Allen T Craig. 2005. *Introduction to mathematical statistics*. Pearson Education.

Pin Huang, Andrew MacKinlay, and Antonio Jimeno Yepes. 2016. Syndromic surveillance using generic medical entities on twitter. In *Proceedings of the Australasian Language Technology Association Workshop 2016*, pages 35–44, Melbourne, Australia.

Hayate Iso, Shoko Wakamiya, and Eiji Aramaki. 2016. Forecasting word model: Twitter-based influenza surveillance and prediction. In *Proceedings of COLING 2016, the 26th International Conference on Computational Linguistics: Technical Papers*, pages 76–86, Osaka, Japan. The COLING 2016 Organizing Committee.

Shin Kanouchi, Mamoru Komachi, Naoaki Okazaki, Eiji Aramaki, and Hiroshi Ishikawa. 2015. Who caught a cold ? - identifying the subject of a symptom. In *Proceedings of the 53rd Annual Meeting of the Association for Computational Linguistics and the 7th International Joint Conference on Natural Language Processing (Volume 1: Long Papers)*, pages 1660–1670, Beijing, China. Association for Computational Linguistics.

Juliana Kaplan, Lauren Frias, and Morgan McFall-Johnsen. 2020. Countries around the world are re-opening — here's our constantly updated list of how they're doing it and who remains under lockdown. *Business Insider*.

Kathy Katella. 2020. Our new covid-19 vocabulary—what does it all mean? *Yale Medicine*.

Yoshiaki Kitagawa, Mamoru Komachi, Eiji Aramaki, Naoaki Okazaki, and Hiroshi Ishikawa. 2015. Disease event detection based on deep modality analysis. In *Proceedings of the ACL-IJCNLP 2015 Student Research Workshop*, pages 28–34, Beijing, China. Association for Computational Linguistics.

Ramez Kouzy, Joseph Abi Jaoude, Afif Kraitem, Molly B El Alam, Basil Karam, Elio Adib, Jabra Zarka, Cindy Traboulsi, Elie W Akl, and Khalil Baddour. 2020. Coronavirus goes viral: quantifying the covid-19 misinformation epidemic on twitter. *Cureus*, 12(3).

Alex Lamb, Michael J. Paul, and Mark Dredze. 2013. Separating fact from fear: Tracking flu infections on twitter. In *Proceedings of the 2013 Conference of the North American Chapter of the Association for Computational Linguistics: Human Language Technologies*, pages 789–795, Atlanta, Georgia. Association for Computational Linguistics.

Nikhil Kumar Rajput, Bhavya Ahuja Grover, and Vipin Kumar Rathi. 2020. Word frequency and sentiment analysis of twitter messages during coronavirus pandemic. *arXiv preprint arXiv:2004.03925*.

Lisa Singh, Shweta Bansal, Leticia Bode, Ceren Budak, Guangqing Chi, Kornraphop Kawintiranon, Colton Padden, Rebecca Vanarsdall, Emily Vraga, and Yanchen Wang. 2020. A first look at covid-19 information and misinformation sharing on twitter. *arXiv preprint arXiv:2003.13907*.

Han Xiao. 2018. bert-as-service. https://github.com/hanxiao/bert-as-service.

COVID-19 and Arabic Twitter: How can Arab World Governments and Public Health Organizations Learn from Social Media?

Lama Alsudias
King Saud University / Saudi Arabia
Lancaster University / UK
`lalsudias@ksu.edu.sa`

Paul Rayson
Lancaster University / UK
`p.rayson@lancaster.ac.uk`

Abstract

In March 2020, the World Health Organization announced the COVID-19 outbreak as a pandemic. Most previous social media related research has been on English tweets and COVID-19. In this study, we collect approximately 1 million Arabic tweets from the Twitter streaming API related to COVID-19. Focussing on outcomes that we believe will be useful for Public Health Organizations, we analyse them in three different ways: identifying the topics discussed during the period, detecting rumours, and predicting the source of the tweets. We use the k-means algorithm for the first goal with k=5. The topics discussed can be grouped as follows: COVID-19 statistics, prayers for God, COVID-19 locations, advise and education for prevention, and advertising. We sample 2000 tweets and label them manually for false information, correct information, and unrelated. Then, we apply three different machine learning algorithms, Logistic Regression, Support Vector Classification, and Naïve Bayes with two sets of features, word frequency approach and word embeddings. We find that Machine Learning classifiers are able to correctly identify the rumour related tweets with 84% accuracy. We also try to predict the source of the rumour related tweets depending on our previous model which is about classifying tweets into five categories: academic, media, government, health professional, and public. Around (60%) of the rumour related tweets are classified as written by health professionals and academics.

1 Introduction

The current coronavirus disease (COVID-19) outbreak is of major global concern and is classified by the World Health Organization as an international health emergency. Governments around the world have taken different decisions in order to stop the spread of the disease. Many academic researchers in various fields including Natural Language Processing (NLP) have carried out studies targeting this subject. For example, COVID-19 and AI[1] is one of the conferences that has been convened virtually to show how Artificial Intelligence (AI) can contribute in helping the Public Health Organizations during pandemics.

People use social media applications such as Twitter to find the news related to COVID-19 and/or express their opinions and feelings about it. As a result, a vast amount of information could be exploited by NLP researchers for a myriad of analyses despite the informal nature of social media writing style.

We hypothesise that Public Health Organizations (PHOs) may benefit from mining the topics discussed between people during the pandemic. This may help in understanding a population's current and changing concerns related to the disease and help to find the best solutions to protect people. In addition, during an outbreak people themselves will search online for reliable and trusted information related to the disease such as prevention and transmission pathways. COVID-19 Twitter conversations may not correlate with the actual disease epidemiology. Therefore, Public Health Organizations have a vested interest in ensuring that information spread in the population is accurate (Vorovchenko et al., 2017). For instance, the Ministry of Health in Saudi Arabia[2] presents a daily press conference incorporating the aim of quickly stopping the spread of false rumours. However, there is currently a prolonged period of time until warnings are issued. For example, the first tweet that included false information about hot weather killing the virus was on 10

---

[1] `https://hai.stanford.edu/events/covid-19-and-ai-virtual-conference`
[2] `https://www.moh.gov.sa`

February 2020, while the press conference which responded to this rumour on 14 April 2020. There is a clearly a need to find false information as quickly as possible. In addition, effort needs to be made in relation to tracking the user accounts that promote rumours. This can be undertaken using a variety of techniques, for example using social network features, geolocation, bot detection or content based approaches such as language style. Public Health Organizations would benefit from speeding up the process of tracking in order to stop rumours and remove the bot networks.

The vast majority of the previous research in this area has been on English Twitter content but this will not directly assist PHOs in Arabic speaking countries. The Arabic language is spoken by 467 million people in the world and has more than 26 dialects[3]. Of particular importance for NLP is coping with dialectal and/or meaning differences in less formal settings such as social media. As an example in health field, the word (حصين) may be understood as vaccination[4] in Modern Standard Arabic or reading supplications in Najdi dialect[5]. There has been much recent progress in Arabic NLP research yet there is still an urgent need to develop fake news detection for Arabic tweets (Mouty and Gazdar, 2018).

In this paper, we have combined qualitative and quantitative studies to analyse Arabic tweets aiming to support Public Health Organizations who can learn from social media data along various lines:

- Analyzing the topics discussed between people during the peak of COVID-19

- Identifying and detecting the rumours related to COVID-19.

- Predicting the type of sources of tweets about COVID-19.

2 Related Work

There is a vast quantity of research over recent years that analyses social media data related to different pandemics such as H1N1 (2009), Ebola (2014), Zika Fever (2015), and Yellow Fever (2016). These studies followed a variety of directions for analysis with multiple different goals (Joshi et al., 2019). The study of Ahmed et al. (2019) used a thematic analysis of tweets related to the H1N1 pandemic. Eight key themes emerged from the analysis: emotion, health related information, general commentary and resources, media and health organisations, politics, country of origin, food, and humour and/or sarcasm.

A survey study (Fung et al., 2016b) reviewed the research relevant to the Ebola virus and social media. It compared research questions, study designs, data collection methods, and analytic methods. Ahmed et al. (2017b) used content analysis to identify the topics discussed on Twitter at the beginning of the 2014 Ebola epidemic in the United States. In (Vorovchenko et al., 2017), they determined the geolocation of the Ebola tweets and named the accounts that interacted more on Twitter related to the 2014 West African Ebola outbreak. The main goal of the study by Kalyanam et al. (2015) was to distinguish between credible and fake tweets. It highlighted the problems of manual labeling process with verification needs. The study in (Fung et al., 2016a) highlighted how the problem of misinformation changed during the disease outbreak and recommended a longitudinal study of information published on social media. Moreover, it pointed out the importance of understanding the source of this information and the process of spreading rumours in order to reduce their impact in the future.

Ghenai and Mejova (2017) tracked Zika Fever misinformation on social media by comparing them with rumours identified by the World Health Organization. Also, they pointed out the importance of credible information sources and encouraged collaboration between researchers and health organizations to rapidly process the misinformation related to health on social media. The study in (Ortiz-Martínez and Jiménez-Arcia, 2017) reviewed the quality of available yellow fever information on Twitter. It also showed the significance of the awareness of misleading information during pandemic spread. The study of Zubiaga et al. (2018a) summarised other studies related to social media rumours. It illustrated techniques for developing rumour detection, rumour tracking, rumour stance classification, and rumour veracity classification. Vorovchenko

---

[3]https://en.wikipedia.org/wiki/Arabic
[4]https://en.wikipedia.org/wiki/Modern_Standard_Arabic
[5]https://en.wikipedia.org/wiki/Najdi_Arabic

et al. (2017) mentioned the importance of Twitter information during the epidemic and how Public Health Organisations can benefit from this. They showed the requirement to monitor false information posted by some accounts and recommended that this was performed in real time to reduce the danger of this information. Also, they discussed the lack of available datasets which help in the development of rumour classification systems.

Researchers have been doing studies on building and analysing COVID-19 Twitter datasets since the disease appeared in December 2019. So far, there are two different datasets which have been published recently related to Arabic and COVID-19 (Alqurashi et al., 2020) and (Haouari et al., 2020). The former collected 3,934,610 million tweets until April 15 2020, and the latter included around 748k tweets until March 31, 2020. These papers contain an initial analysis and statistical results for the collected tweets and some suggestions for future work, which include pandemic response, behavior analysis, emergency management, misinformation detection, and social analytics.
On the other hand, there are some datasets in English such as (Chen et al., 2020) and (Lopez et al., 2020). Also, there is a multilingual COVID-19 dataset containing location information (Qazi et al., 2020). This contains more than 524 million tweets, with 5.5 million Arabic tweets, posted over a period of three months since February 1, 2020. It focuses on determining the geolocation of a tweet which can help research with various different challenges, including identifying rumours.

Although the above studies have produced datasets related to COVID-19, they do not analyse them deeply using NLP methods. Previous studies representing earlier epidemics present good techniques and results, however none of them are related to Arabic tweets. Therefore, to assist PHOs in Arabic speaking countries there is an urgent need to analyse tweets related to COVID-19 using multiple Arabic NLP techniques.

3   Update Arabic Infectious Disease Ontology

With the recent appearance of COVID-19 as a new disease, there is need to update our Arabic Infectious Disease Ontology (Alsudias and Rayson, 2020), which integrates the scientific and medical vocabularies of infectious diseases with their informal equivalents used in general discourse. We collated COVID-19 information from the World Health Organization[6] and Ministry of health in Saudi Arabia. This included symptom, cause, prevention, infection, organ, treatment, diagnosis, place of the disease spread, and slang terms for COVID-19 and extended our ontology[7]. These terms were then used in our collection process.

4   Data Collection

We began collecting Arabic tweets about a number of infectious diseases from September 2019. Here in this paper, we analysed only the tweets related to COVID-19 from December 2019 to April 2020 (there are a few tweets between September and November, these are related to Middle East respiratory syndrome coronavirus, MERS-CoV[8]). We have collected approximately six million tweets in Arabic during this period. We obtained the tweets depending on three keywords (كورونَا, كرونَا, كوفِيد ١٩) which mean Coronavirus, a misspelling of the name of Coronavirus, and COVID-19 respectively in English. We collected the tweets weekly using Twitter API.

Next, we pre-processed the tweets through a pipeline of different steps:

- Manually remove retweets, advertisements, and spam.

- Filter out URLs, mentions, hashtags, numbers, emojis, repeating characters, and non-Arabic words using Python scripts[9].

- Normalize and tokenize tweets.

- Remove Arabic stopwords (Alrefaie, 2017).

After pre-processing, the resulting dataset was 1,048,575 unique tweets from the original 6,578,982 collected. Figure 1 shows the number of Arabic tweets about Coronavirus each week with specific dates highlighted to show government decisions on protecting the population from COVID-19 and other key dates for context.

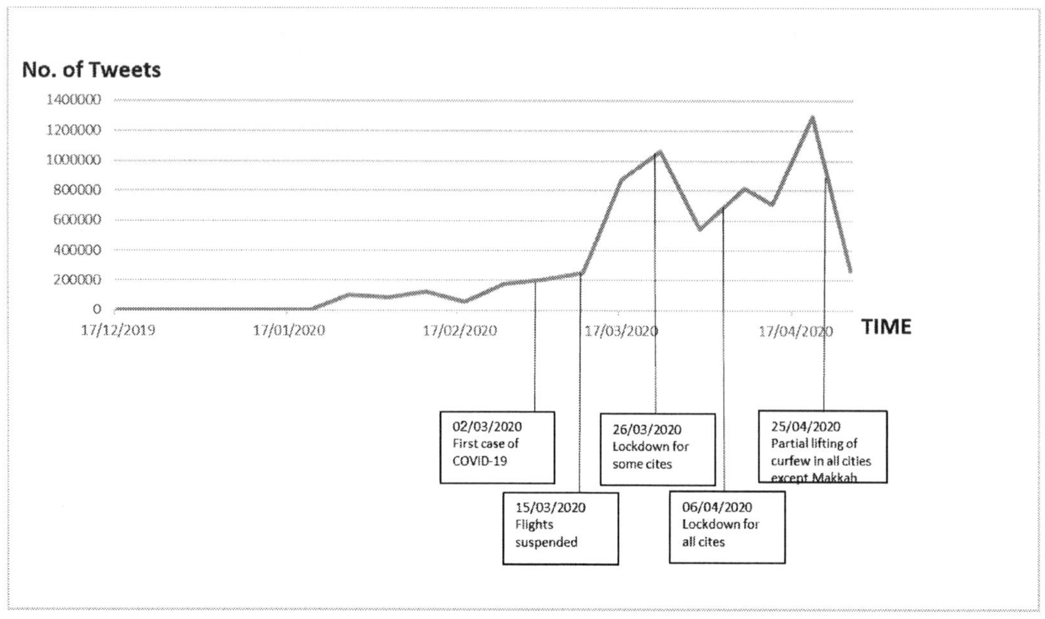

Figure 1: Number of Arabic tweets about Coronavirus

5 Methods

We performed three different types of analysis on the collected data. Firstly, in order to better understand the topics discussed in the corpus, we carried out a cluster analysis. Secondly, taking a sample of the corpus, we performed rumour detection. Finally, we extended our previous work to classify the source of tweets into five types of Twitter users which aims at helping to determine their veracity.

5.1 Cluster Analysis

To explore the topics discussed on Twitter during the COVID-19 epidemic in Saudi Arabia and other countries in the Arab World, we subjected the text of the tweets to cluster analysis. After pre-processing the tweets as described above, we used the N-gram forms (unigram, bigram, and tri-gram) of twitter corpus and clustered them using the K-means algorithm with the Python Scikit-learn 0.20.2 (Pedregosa et al., 2011) software and set the value of k, the number of clusters, to be five.

5.2 Rumour Detection

Following previous work (Zubiaga et al., 2018b), we applied a top-down strategy, which is where the set of rumours is identified in advance then the data is sampled to extract the posts associated with the previously identified rumours. In our dataset, out of the one million tweets, we sampled 2,000 tweets to classify them for rumour detection. We manually labelled the tweets to create a gold standard dataset and then applied different machine learning algorithms in this part of our study.

5.2.1 Labelling Guidelines

We manually labelled the tweets with 1, -1 , and 0 to represent false information, correct information, and unrelated content, respectively. Our reference point for deciding whether the content contained true or false information was based on the list issued by the Ministry of Health in Saudi Arabia [10] and is regularly updated (the last update applied for this study dates from 14 April 2020). Table 1 presents the list in both Arabic and English and Table 2 shows some example tweets for each label.

5.2.2 Machine Learning Models

We applied three different machine learning algorithms: Logistic Regression (LR), Support Vector Classification (SVC), and Naïve Bayes (NB). To help the classifier distinguish between the classes more accurately, we extracted further linguistic features. The selected features fall into two groups: word frequency, count vector and TF-IDF, and word embedding based (Word2Vec and FastText). We used 10-fold cross validation to determine accuracy of the classifiers for this dataset, splitting the

```
middle-east-respiratory-syndrome-coronavirus-(mers-cov)
```
[9]`https://github.com/alsudias`   [10]`https://www.moh.gov.sa`

Rumour in Arabic	Rumour in English
الحيوَانَات الَاليفة تنقل فيروس كورونَا.	Pets are transporters of Coronavirus.
البعوض نَاقل لّكورونَا.	Mosquitoes are transporters of Coronavirus.
الَاطفَال قد لَا يصَابون بفيروس كورونَا.	Children are not infected by Coronavirus.
كبَار السن فقط قد يتعرضون لمخَاطر سيئة من فيروس كورونَا.	Only old people may have a high risk of Coronavirus.
حرَارة الطقس أو برودته تقضي علَى الفيروس.	Hot or cold weather can kill the virus.
الغرغرة بَالمَاء و الملح تقضي علَى الفيروس.	Gargling with water and salt eliminates the virus.
يوجد خلطَات و أعشَاب تقي من الكورونَا.	There are some herbs that protect against from Coronavirus.
الفيروس لَا يبقَى علَى الَاسطح.	The virus does not survive on surfaces.

Table 1: List of rumours that appear during COVID-19 (source: Saudi Arabia Ministry of Health)

Tweet in Arabic	Tweet in English	Label
سيكون هنَاك انحسَار لَانتشَار فيروس كورونَا مع بدَاية فصل الصيف خصوصًا في العَالم العربي نظرًا لَارتفَاع درجة الحرَارة.	There will be a decrease in the spread of the Corona virus at the beginning of the summer, especially in the Arab world, due to the high temperatures.	1 (false)
الصحة: يعيش الفيروس و يرتكز بَالَاسَاس في الجهَاز التنفسي لذلك غير وَارد انتقَاله عن طريق الحشرَات أو من خلَال لدغة البعوض.	The Ministry of Health: A virus lives and is mainly concentrated in the respiratory system, so it is not likely to be transmitted by insects or by mosquito bites.	-1 (true)
اللّهم في هذي السَاعة المبَاركة نسآلك ان ترحمنَا و تبعد عنَا كل دَاء و بلَاء و قنَا شر الَامرَاض و الَاسقَام. وَاحفظ بلَادنَا و كَافة بلَاد المسلمين.	Oh God, in this blessed hour, We ask you to have mercy on us and keep away from us all disease and calamity, and protect us from the evil of diseases and sicknesses. Preserve our country and other Muslim countries.	0 (unrelated)

Table 2: Example tweets and our labelling system

entire sample into 90% training and 10% testing for each fold.

5.3 Source Type Prediction

We replicated a Logic Regression model from our previous study, which was useful for classifying tweets into five categories: academic, media, government, health professional, and public (Alsudias and Rayson, 2019). We used this LR model because it previously achieved the best accuracy (77%), and employed it here to predict the source of the COVID-19 tweets that we had already labelled in Section 5.2.

6 Results and Discussion

6.1 Cluster Analysis

Our cluster analysis of the five main public topics discussed in tweets content is as follows: (1)

disease statistics: the number of infected, died, and recovered people; (2) prayers: prayer asking God to stop virus; (3) disease locations: spread and location (i.e., name of locations, information about spread); (4) Advise for prevention education: health information (i.e., prevention methods, signs, symptoms); and (5) advertising: adverts for any product either related or not related to the virus. Figure 2 illustrates the five topics of COVID-19 tweets with examples.

For each cluster, the top terms by frequency are as follows: (1) disease statistics: case (حَالة), new (جديدة), and infection (أصَابة); (2) prayers: Allah (الله), Oh God (اللّهم), and Muslims (المسلمين); (3) disease locations: Dammam (الدمَام), Riyadh (الريَاض), and Makkah (مكة); (4) Advise for prevention education: crisis (أزمة), spread (انتشَار), and pandemic (جَائِحة); (5) advertising: discount (خصم), coupon (كوبون), and code (كود).

We found that four of our categories (disease statistics, prayers, disease locations, and advise for prevention education) are similar to those found by Odlum and Yoon (2015) which are risk factors, prevention education, disease trends, and compassion. The marketing category is one of the topics in (Ahmed et al., 2017a) which discussed the topics in Twitter during the Ebola epidemic in the United States. Jokes and/or sarcasm is one of the categories that did not appear in our study but can be found in (Ahmed et al., 2017a) and (Ahmed et al., 2019), a thematic analysis study of Twitter data during H1N1 pandemic. This may be a result of more concern and panic from COVID-19 than other diseases during this period of time.

6.2 Rumour Detection

The result of our manual labelling process is 316 tweets label with 1 (false), 895 tweets label with 1 (true), and 789 tweets label with 0 (unrelated). Therefore, the false information represents about 15.8% (from 2,000 tweets) and around 26% (from 1,211 tweets, after removing the unrelated ones). In the study by Ortiz-Martínez and Jiménez-Arcia (2017), 61.3% (from 377 tweets) of data was classified as misinformation about Yellow Fever. It represented 32% (from 26,728 tweets) considered as rumours related to Zika Fever in Ghenai and

Mejova (2017).

Figures 3, 4, and 5 show the accuracy, F1-score, recall, and precision on our corpus using LR, SVC, and NB algorithms with various feature selection approaches. The highest accuracy (84.03%) was achieved by the LR classifier with a count vector set of features and SVC with TF-IDF. Therefore, the count vector gives best result in LG and NB in all metrics results whereas with precision which achieves better results with TF-IDF 83.71% in LG and 81.28% in NB. while TF-IDF in SVC has the best results except recall which achieved 75.55% in count vector set features.

We also applied several word embedding based approaches but without obtaining good results. The accuracy ranges from 50% to 60% and the F1 score is around 40% on average. FastText models achieve better accuracy in SVC (54.89%) and NB (59.49%) than Word2Vec by approximately (5%). While Word2Vec shows the best result with LG 60.68% for accuracy, 49% for F1 and recall, and 65.97% in precision.

The word frequency based approaches have around a 20% better result than the word embedding ones. The reason for this is expected to be the dataset size and the specific domain of context (Ma, 2018). We assumed that the word embedding methods may achieve good results due to the importance of the relevant information around the word. For example, FastText can deal with the misspelling problem which is common in social media language style and improves word vectors with subword information (Bojanowski et al., 2017).

6.3 Source Type Prediction

The model predicts the source type for each of the tweets. Table 3 shows some examples of the tweets with predicted labels by the model. We focus on the result of the fake news content since they are of highest importance for the Public Health Organization. 30% (95 of 316) and 28% (91 of 316) of the rumour tweets are classified as written by a health professional and academic consequently. While only 12% (39 of 316) of them are predicted as written by the public. With this result, we find that the tweets containing false information quite often used the language style of academics and health professionals.

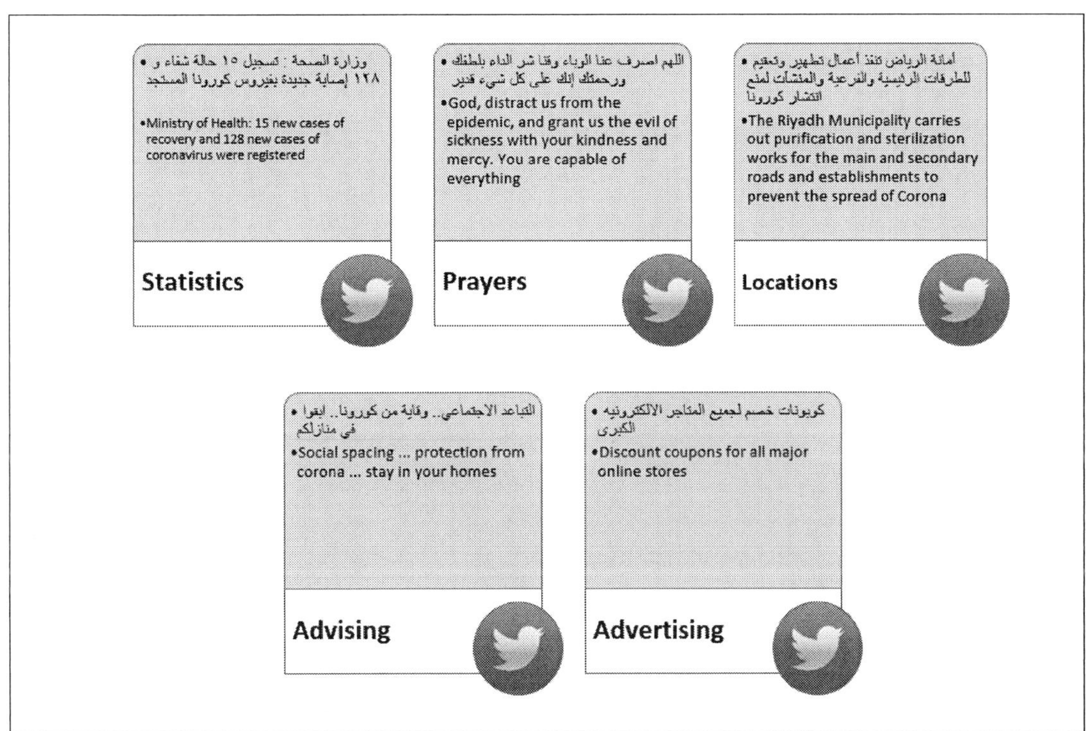

Figure 2: Examples of tweets in each cluster

Tweet in Arabic	Tweet in English	Predicted Label
في قراءة علمية... يتوقع اندثار الفيروس في ابريل بسبّب الحرارة	In scientific reading ... the virus is expected to erode in April due to heat.	Academic
متحدث الصحة الحالات المؤكدة حتّى الآن المصابة بفيروس كورونا و معضمها لبالغين	Health spokesman has confirmed cases so far infected with coronavirus, mostly for adults.	Media
يا وزارة رشوا البعوض ، هو الناقل لفيروس كورونا زادت الاصابات مع انتشار البعوض	Ministry of health please spray mosquitoes, as they are carriers of the Coronavirus, increased infections as mosquitoes spread.	Government
خبير صيني يؤكد ان استنشاق بخار الماء يقتل فيروس كورونا	A Chinese expert confirms that inhaling water vapor kills coronavirus.	Health professional
علاج الكرونا بالليمون و الثوم "رابط يوتيوب"	Corona treatment with lemon and garlic "YouTube link".	Public

Table 3: Some examples of false tweets from different source predicted labels

7 Conclusion and Future work

In this paper, we identified and analysed one million tweets related to the COVID-19 pandemic in the Arabic language. We performed three experiments which we expect can help to develop methods of analysis suitable for helping Arab World Governments and Public Health Organisations. Our analysis first identifies the topics discussed on social media during the epidemic, detects the tweets that contain false information,

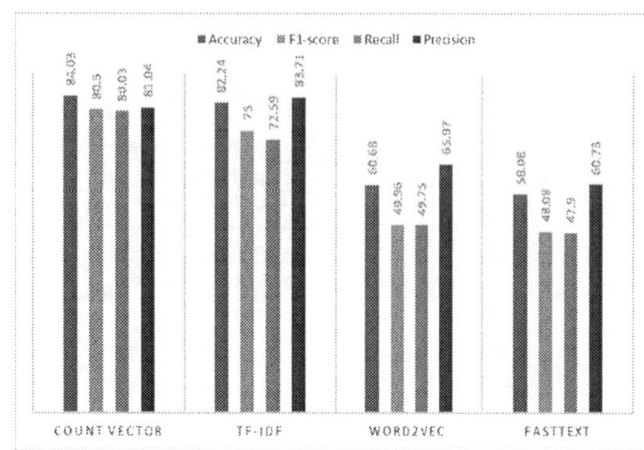

Figure 3: Results using Logistic Regression

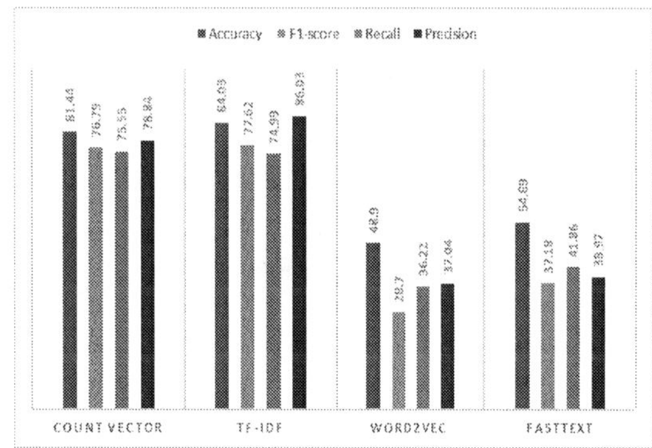

Figure 4: Results using Support Vector Classification

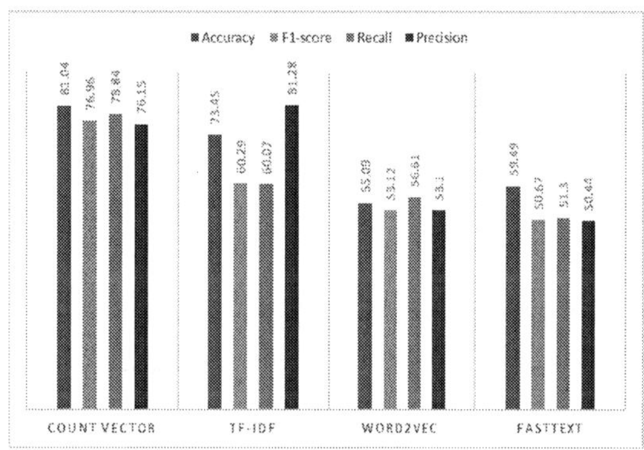

Figure 5: Results using Naïve Bayes

and predicts the source of the rumour related tweets based on our previous model for other diseases. The clustered topics are COVID-19 statistics, prayers for God, COVID-19 locations, advise for preventing education, and advertising.

Our second contribution is a labeled sample of tweets (2,000 out of 1 million) annotated for false information, correct information, and unrelated. To investigate the replicability and scalability of this annotation, we applied multiple Machine Learning Algorithms with different sets of features. The highest accuracy result was 84% achieved by the LR classifier with count vector set of features and SVC with TF-IDF.

Finally, we also used our previous model to predict the source types of the sampled tweets. Around 60% of the rumour related tweets are classified as written by health professional and academics which shows the urgent need to respond to such fake news. The dataset, including tweet IDs, manually assigned labels for the sampled tweets, and other resources used in this paper are made freely available for academic research purposes [11].

There are clearly many potential future directions related to analysing social media data on the topics of pandemics. Since false information has the potential to play a dangerous role in topics related to health, there is a need to enhance and automate the automatic detection process supporting different languages beyond just English. Future potential directions include monitoring the spread of the disease by finding the infected individuals, defining the infected locations, or observing people that do not apply self isolation rules. Moreover, the analysis could proceed in an exploratory and thematic way such as discovering further topics discussed during the epidemic, as well as assisting governments and public health organisations in measuring people's concerns resulting from the disease.

References

Wasim Ahmed, Peter A Bath, Laura Sbaffi, and Gianluca Demartini. 2019. Novel insights into views towards H1N1 during the 2009 Pandemic: a thematic analysis of Twitter data. *Health Information & Libraries Journal*, 36(1):60–72.

Wasim Ahmed, Gianluca Demaerini, and Peter A Bath. 2017a. Topics discussed on twitter at the beginning of the 2014 Ebola epidemic in United States. *iConference 2017 Proceedings*.

Wasim Ahmed, G. Demartini, and P. Bath. 2017b. Topics Discussed on Twitter at the Beginning of the 2014 Ebola Epidemic in United States.

Sarah Alqurashi, Ahmad Alhindi, and Eisa Alanazi. 2020. Large arabic twitter dataset on covid-19. *arXiv preprint arXiv:2004.04315*.

[11] https://doi.org/10.17635/lancaster/researchdata/375

Mohamed Taher Alrefaie. 2017. arabic stop words. https://github.com/mohataher/arabic-stop-words.

Lama Alsudias and Paul Rayson. 2019. Classifying information sources in Arabic twitter to support online monitoring of infectious diseases. In *Proceedings of the 3rd Workshop on Arabic Corpus Linguistics*, pages 22–30, Cardiff, United Kingdom. Association for Computational Linguistics.

Lama Alsudias and Paul Rayson. 2020. Developing an Arabic Infectious Disease Ontology to Include Non-Standard Terminology. In *Proceedings of The 12th Language Resources and Evaluation Conference*, pages 4844–4852, Marseille, France. European Language Resources Association.

Piotr Bojanowski, Edouard Grave, Armand Joulin, and Tomas Mikolov. 2017. Enriching word vectors with subword information. *Transactions of the Association for Computational Linguistics*, 5:135–146.

Emily Chen, Kristina Lerman, and Emilio Ferrara. 2020. Covid-19: The first public coronavirus twitter dataset. *arXiv preprint arXiv:2003.07372*.

I.C.-H Fung, King-wa Fu, C.-H Chan, Benedict Chan, Chi-Ngai Cheung, Thomas Abraham, and Zion Tse. 2016a. Social Media's Initial Reaction to Information and Misinformation on Ebola, August 2014: Facts and Rumors. *Public Health Reports*, 131:461–473.

Isaac Chun-Hai Fung, Carmen Hope Duke, Kathryn Cameron Finch, Kassandra Renee Snook, Pei-Ling Tseng, Ana Cristina Hernandez, Manoj Gambhir, King-Wa Fu, and Zion Tsz Ho Tse. 2016b. Ebola virus disease and social media: A systematic review. *American journal of infection control*, 44(12):1660—1671.

Amira Ghenai and Yelena Mejova. 2017. Catching Zika fever: Application of crowdsourcing and machine learning for tracking health misinformation on Twitter. *arXiv preprint arXiv:1707.03778*.

Fatima Haouari, Maram Hasanain, Reem Suwaileh, and Tamer Elsayed. 2020. The First Arabic COVID-19 Twitter Dataset with Propagation Networks. *arXiv preprint arXiv:2004.05861*.

Aditya Joshi, Sarvnaz Karimi, Ross Sparks, Cécile Paris, and C Raina Macintyre. 2019. Survey of Text-based Epidemic Intelligence: A Computational Linguistics Perspective. *ACM Computing Surveys (CSUR)*, 52(6):1–19.

Janani Kalyanam, Sumithra Velupillai, Son Doan, Mike Conway, and Gert R. G. Lanckriet. 2015. Facts and Fabrications about Ebola: A Twitter Based Study. *ArXiv*, abs/1508.02079.

Christian E Lopez, Malolan Vasu, and Caleb Gallemore. 2020. Understanding the perception of COVID-19 policies by mining a multilanguage Twitter dataset. *arXiv preprint arXiv:2003.10359*.

Edward Ma. 2018. 3 basic approaches in Bag of Words which are better than Word Embeddings.

Rabeaa Mouty and Achraf Gazdar. 2018. Survey on Steps of Truth Detection on Arabic Tweets. In *2018 21st Saudi Computer Society National Computer Conference (NCC)*, pages 1–6. IEEE.

Michelle Odlum and Sunmoo Yoon. 2015. What can we learn about the Ebola outbreak from tweets? *American journal of infection control*, 43(6):563–571.

Yeimer Ortiz-Martínez and Luisa F Jiménez-Arcia. 2017. Yellow fever outbreaks and Twitter: Rumors and misinformation. *American journal of infection control*, 45(7):816–817.

Fabian Pedregosa, Gaël. Varoquaux, Alexandre Gramfort, Vincent Michel, Bertrand Thirion, Olivier Grisel, Mathieu Blondel, Peter Prettenhofer, Ron Weiss, Vincent Dubourg, Jake Vanderplas, Alexandre Passos, David Cournapeau, Matthieu Brucher, Matthieu Perrot, and Édouard Duchesnay. 2011. Scikit-learn: Machine Learning in Python. *Journal of Machine Learning Research*, 12:2825–2830.

Umair Qazi, Muhammad Imran, and Ferda Ofli. 2020. GeoCoV19: A Dataset of Hundreds of Millions of Multilingual COVID-19 Tweets with Location Information. *arXiv preprint arXiv:2005.11177*.

Tatiana Vorovchenko, Proochista Ariana, Francois van Loggerenberg, and Amirian Pouria. 2017. *Ebola and Twitter. What Insights Can Global Health Draw from Social Media?*, pages 85–98.

Arkaitz Zubiaga, Ahmet Aker, Kalina Bontcheva, Maria Liakata, and Rob Procter. 2018a. Detection and Resolution of Rumours in Social Media: A Survey. *ACM Comput. Surv.*, 51(2).

Arkaitz Zubiaga, Ahmet Aker, Kalina Bontcheva, Maria Liakata, and Rob Procter. 2018b. Detection and resolution of rumours in social media: A survey. *ACM Computing Surveys (CSUR)*, 51(2):1–36.

NLP-based Feature Extraction for the Detection of COVID-19 Misinformation Videos on YouTube

Juan Carlos Medina Serrano, Orestis Papakyriakopoulos, Simon Hegelich

Technical University of Munich, Germany

{juan.medina, orestis.p}@tum.de, simon.hegelich@hfp.tum.de

Abstract

We present a simple NLP methodology for detecting COVID-19 misinformation videos on YouTube by leveraging user comments. We use transfer learning pre-trained models to generate a multi-label classifier that can categorize conspiratorial content. We use the percentage of misinformation comments on each video as a new feature for video classification. We show that the inclusion of this feature in simple models yields an accuracy of up to 82.2%. Furthermore, we verify the significance of the feature by performing a Bayesian analysis. Finally, we show that adding the first hundred comments as tf-idf features increases the video classifier accuracy by up to 89.4%.

1 Introduction

The COVID-19 health crisis was accompanied by a misinfodemic: The limited knowledge on the nature and origin of the virus gave ample space for the emergence of conspiracy theories, which were diffused on YouTube, and online social networks. Although YouTube accelerated attempts to detect and filter related misinformation, it yielded moderate results (Li et al., 2020; Frenkel et al., 2020).

In this study, we present a simple NLP-based methodology that can support fact checkers in detecting COVID-19 misinformation on YouTube. Instead of training models on the videos themselves and predicting their nature, we exploit the vast amount of available comments on each YouTube video and extract features that can be used in misinformation detection. Our methodology comes with the advantage that labeling comments is simpler and faster than video labeling. Additionally, no complex neural architecture is needed for the classification of videos.

Our study provides the following contributions:

- We create a multi-label classifier based on transfer learning that can detect conspiracy-laden comments. We find that misinformation videos contain a significantly higher proportion of conspiratorial comments.

- Based on this information, we use the percentage of conspiracy comments as feature for the detection of COVID-19 misinformation videos. We verify its efficiency by deploying simple machine learning models for misinformation detection. We employ the videos' title and the first 100 comments to validate feature significance.

- We show that including the first hundred comments as tf-idf features in the classifier increases accuracy from 82.2% to 89.4%.

2 Related Work

Previous research studies have extensively investigated the possibilities and limits of NLP for detecting misinformation. Researchers have provided theoretical frameworks for understanding the lingual and contextual properties of various types of misinformation, such as rumors, false news, and propaganda (Li et al., 2019; Thorne and Vlachos, 2018; Rubin et al.; Zhou and Zafarani, 2018). Given the general difficulty in detecting misinformation, scientists have also developed dedicated benchmark datasets to evaluate the effectiveness of NLP architectures in misinformation-related classification tasks (Pérez-Rosas et al., 2018; Hanselowski et al., 2018). Given the vast amount of misinformation appearing in online social networks, various research studies propose case-specific NLP methodologies for tracing misinformation. For example, Della Vedova et al. (2018) and Popat et al. (2018) combined lingual properties of articles and other meta-data for the detection of false news. Volkova et al. (2017), Qazvinian et al. (2011) and Kumar and Carley (2019) created special architectures that take into consideration the microblogging structure of online social networks, while De Sarkar et al. (2018) and Gupta et al. (2019) exploited sentence-level semantics for misinformation detection.

Despite the deployment of such architectures for fact checking, locating malicious content and promptly removing it remains an open challenge (Gillespie, 2018; Roberts, 2019). In the case of COVID-19 misinformation, a large share of conspiratorial contents remain online on YouTube and other platforms, influencing the public despite content moderation practices (Li et al., 2020; Frenkel et al., 2020; Ferrara, 2020). Given this, it is important to develop case-specific NLP tools that can assist policymakers and researchers in the process of detecting COVID-19 misinformation and managing it accordingly. Towards this end, we illustrate how NLP-based feature extraction (Shu et al., 2017; Jiang et al., 2020) based on user comments can be effectively used for this task. User comment data has been employed to annotate social media objects (Momeni et al., 2013), infer the political leaning of news articles (Park et al., 2011), and to predict popularity (Kim et al., 2016). Previous studies explicitly employed comments as proxies for video content classification (Huang et al., 2010; Filippova and Hall, 2011; Eickhoff et al., 2013; Doğruöz et al., 2017). However, only Jiang and Wilson (2018) have analyzed user content to identify misinformation. However, they focused on linguistic signals and concluded that users' comments were not strong signals for detecting misinformation.

3 Methodology and Experiments

3.1 Dataset

The first step of the study consisted of obtaining a set of YouTube videos that included either misinformation or debunking content. We decided to search for YouTube videos through user-generated content on social media platforms. For this, we queried the Pushshift Reddit API (Baumgartner et al., 2020), and Crowdtangle's historical data of public Facebook posts (Silverman, 2019) using the query "COVID-19 OR coronavirus". Additionally, we downloaded the COVID-19 Twitter dataset developed by Chen et al. (2020). The total dataset included over 85 million posts generated between January and April 2020. We significantly reduced this dataset by querying the posts with "biowarfare OR biological weapon OR bioweapon OR man-

made OR human origin". From the remaining posts, we extracted and expanded the URLs. We identified 1,672 unique YouTube videos. 10% of these videos had been blocked by YouTube as of April 2020. For the rest of the videos, we watched them, excluded the non-English videos, and manually labeled them as either misinformation, factual, or neither. To label a video as misinformation, we validated that its message was conveying with certainty a conspiracy theory regarding the origin of the coronavirus, as a man-made bioweapon or being caused by 5G. We did not classify videos that questioned its origin but showed no certainty about a hoax (which included well-known and verified news media videos) as misinformation. We classified as factual those videos that included debunking of conspiracy theories or presented scientific results on the origins and causes of COVID-19. We labeled the rest of the videos as neither. Two of the authors (JCMS, OP) performed the labeling procedure independently. For the cases where the labels did not agree, the third author was consulted (SH).

Afterward, we collected the comments on both misinformation and factual videos using YouTube's Data API[1]. For this study, we only included videos with more than twenty comments. The final dataset consisted of 113 misinformation and 67 factual videos, with 32,273 and 119,294 total comments respectively. We selected a ten percent random sample of the comments from the misinformation videos and proceeded to label them. This labeling procedure was performed in the same manner as the video classification to ensure data quality. For each comment, we collected two labels. First, we gave a label if the comment expressed agreement (1) or not (0). Agreement comments included comments such as "this is the video I was looking for", or "save and share this video before YouTube puts it down". The second label considered if comments amplified misinformation with a conspiracy theory/misinformation comment (1) or without one (0). Comments that questioned the conspiracies (such as "could it be a bioweapon?") were not labeled as misinformation. 19.7% of the comments in the sample were labeled as conspiracy comment and 12.5% as agreement comment. Only 2.2% of the comments were classified as both agreement and conspiratorial. Although both agreement and conspiracy labeled comments express the same message of believing in the misinformation

---

[1] https://developers.google.com/youtube/v3

content from the videos, we decided to keep them apart due to their different linguistic properties. To compare the collection of agree-labeled comments and conspiracy-labeled comments, we tokenized and created a bag-of-words model. 19.4% of the processed tokens appear on both collections. However, only 1.95% of the tokens have more than four occurrences in the two collections. We applied χ^2 tests for each of these remaining words and observed that 50% occur in significantly different proportions. In the end, only 0.96% of the vocabulary has a significant similar number of occurrences in the two datasets. The YouTube comments dataset without user data can be accessed in this GitHub repository[2], alongside a Google Colab notebook with the code.

3.2 Classification of Users' Comments

We first performed a multi-label classification on the 10% sample of the misinformation videos' comments. We split the annotated data into training (80%) and test (20%) datasets. We employed state-of-the-art neural transfer learning for the classification by fine-tuning three pre-trained models: XLNet base (Yang et al., 2019), BERT base (Devlin et al., 2018) and RoBERTa base (Liu et al., 2019). The fine-tuning consists of initializing the model's pre-trained weights and re-training on labeled data. We ran the models for four epochs using the same hyperparameters as the base models. For the experiments, we used 0.5 as a decision threshold. Additionally, we trained two simpler models as baselines: a logistic regression model using LIWC's lexicon-derived frequencies (Tausczik and Pennebaker, 2010) as features, and a multinomial Naive Bayes model using bag-of-words vectors as features. Table 1 shows the average micro-F_1 scores for the three transformer models after performing the fine-tuning five times. RoBERTa

---

[2]https://github.com/JuanCarlosCSE/YouTube_misinfo

	Agree		Conspiracy	
	Train	Test	Train	Test
LIWC	88.7	88.6	81	78.2
NB	94.2	82.4	94.3	78.8
XLNet	97±0.1	93.1±0.3	93.9±0.5	84.8±0.6
BERT	**98.5±0.1**	93.3±0.5	96.3±0.3	83.8±0.9
RoBERTa	98.1±0.2	**93.9±0.4**	**96.4±0.3**	**86.7±0.5**

Table 1: Train and test micro F_1 scores (mean and standard deviation) from multi-label classification models: LIWC with logistic regression and Naive Bayes as baselines, and three transformer models with five runs.

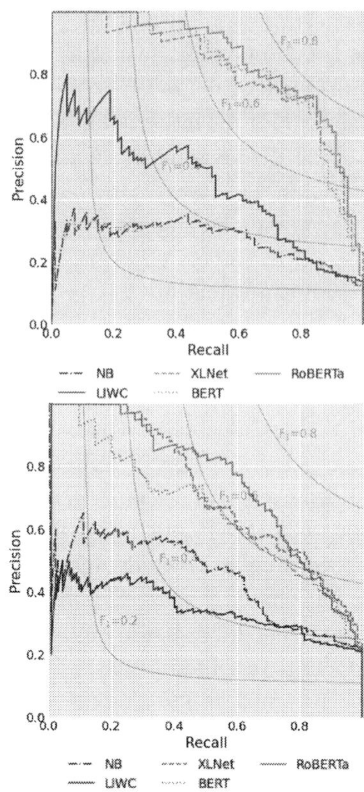

Figure 1: Precision and recall curves for binary F_1 scores for the conspiracy (upper figure) and agreement (lower figure) label. The plot shows the results for three neural-transfer classifiers.

is the best performing model for the training and test dataset on the conspiracy classification as for the test data on the agreement label. BERT is the best performing model only for the training data on the agree label. The three transformer models outperform the baseline models. This predictive superiority is more evident in the precision-recall curves (with corresponding binary-F_1 scores) of the five models on the test data (Figure 1).

We employed the fine-tuned RoBERTa model to predict the labels of the remaining comments from the misinformation and factual videos. We then calculated the *percentage of conspiracy comments* per video. We also obtained this percentage for the agreement label. Figure 2 shows the resulting density distributions from misinformation and factual videos. We observed a difference between the distributions from the two types of videos. We confirmed this by performing Welch's t-test for independent samples. For the conspiracy comments percentage, the t-test was significant (p<0.000), indicating that the samples came from different distributions. The t-test was not significant for the agreement percentage (p>0.1).

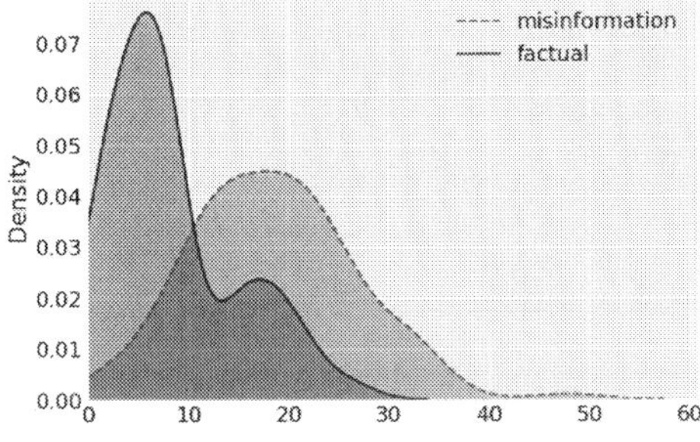

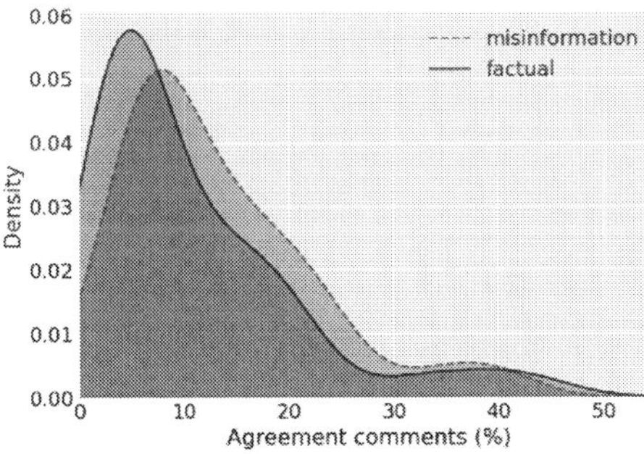

Figure 2: Probability densities of misinformation and factual videos regarding the percentage of conspiratorial comments (top) agreement comments (bottom).

3.3 Classification of YouTube Videos

The next step consisted of classifying the set of YouTube videos to detect misinformation. For this, we employed the percentage of conspiracy comments of each video as a feature. Additionally, we extracted content features from the videos' titles and from the raw first hundred comments per video (or all the comments for videos with fewer than 100 comments). For this, we preprocessed the titles and comments with tokenization, removal of stopwords, and usage of the standard term frequency-inverse document (tf-idf) weighting for word frequencies to create a document term matrix, whose columns serve as input features. We selected six feature settings for our experiments: each of the set of features alone and the three possible combination between them . For each setting, we employed three classification models: logistic regression, linear support vector machine (SVM), and random forest. We performed 10-fold cross-validation and report the mean accuracy in Table 2. We avoided grid search to find better hyperparameters as we did not have a test dataset. We observe that the SVM model has the highest accuracy for all the settings except for one. The conspiracy feature

	LR	SVM	RF
title	62.7	62.7	62.7
conspiracy %	62.7	**81.1**	72.2
comments	66.7	**83.9**	82.8
title + conspiracy %	64.4	77.7	**82.2**
comments + conspiracy %	73.3	**89.4**	84.44
all	73.3	**84.4**	82.7

Table 2: Classification accuracy for logistic regression, linear support vector machines, and random forest models for six feature settings. Results show the average of 10-k cross-validation.

alone achieves an accuracy of 81.1. Using the tf-idf comment features the accuracy is slightly better with 83.9. However, the conspiracy feature and comments combined achieve the highest accuracy of 89.4. We observe that the models with all the features combined have lower accuracy than the models omitting the title features. This may explain that the title is not a good feature. Using the title feature alone does not improve the baseline accuracy of 62.7. Interestingly, the accuracy for the best model is still high (85.5%) when taking into consideration only videos with less than 100 comments. This implies that our methodology is appropriate for the early detection of misinformation videos.

3.4 Bayesian Modeling

To find the statistical validity of the conspiracy percentage feature, we turned to Bayesian modeling as it allows us to obtain the full posterior distribution of feature coefficients. We performed inference on three Bayesian logistic regression models using a Hamiltonian Monte Carlo solver. A simple model considered only the conspiracy percentage feature. A second model included this feature and the ten most relevant word features from the random forest model trained only on the title and conspiracy percentage. A third model included the conspiracy feature, and the top ten most relevant words from the linear SVM trained on the conspiracy feature and the first 100 comments. The first column of Table 3 and 4 shows the importance of each of the features in the random forest and linear SVM model, respectively. The two tables also show the statistics of the posterior probability distributions of the model coefficients: the mean, standard deviation, and the 1% and 99% quantiles. For the three models, the coefficients distribution converged (the $\widehat{R}$ diagnostic (Vehtari et al., 2019) was equal to one). We specifically selected logistic regression models for their interpretability. We observe that for the model based on the title word features, the posterior distribution of the conspiracy percentage feature coefficient is the only one that does not include zero in its 98% highest posterior density interval (Table 3). Although this is not equivalent to traditional p-values, it conveys significance in a Bayesian setting. The model based on the 100 comments word features (Table 4) maintains the conspiracy feature as significant. However, three coefficients from the word features also avoid zero in their 98% interval. The model's coefficients are negative for *covid19* and *lab*, and positive for *god*.

Finally, we compare the three Bayesian models using the WAIC information criteria, which estimates out-of-sample expectation and corrects for the effective number of parameters to avoid overfitting (Watanabe and Opper, 2010). Figure 3 shows the resulting deviance of the three models. We observe that the second model is slightly better than the simple model. However, the differences are included in the standard error of the title words feature model. This is not true for the simple model and the model including the comments features. In this case, the full model outperforms the model based only on the conspiracy feature. This indicates that there is important information in the videos' first hundred comments that is not explained by the conspiracy percentage feature on its own.

4 Discussion

We have leveraged large quantities of user comments to extract a simple feature that is effective in predicting misinformation videos. Given that the classifier is also accurate for videos with few comments, it can be used for online learning. For example, the user comments of videos containing *coronavirus* can be tracked and classified as they are posted. High levels of conspiracy comments could then indicate that the video includes misinformation claims. For this to work, it is not necessary to have a conspiracy classifier with perfect accuracy given that the percentage of conspiracy comments feature is based on aggregating the classification results from all the comments. An improved classifier would be able to define a threshold that allows a balanced number of false positives and true negatives. The average percentage of conspiratorial comments would be maintained, irrespective of the wrong classifications. On the other hand, the accuracy of the video classifier is more critical. We found that using simple classifiers on the raw

	RF	**mean**	**SD**	**1%**	**99%**
conspiracy %	19.2	**28.25**	4.8	18.19	39.94
coronavirus	2.95	-7.45	3.4	-15.57	0.01
covid19	2.81	-5.17	2.4	-11.08	0.10
china	1.42	-4.28	3	-11.23	2.63
man	1.24	-6.04	2.8	-12.25	0.52
bioweapon	1.24	4.81	5.5	-6.40	19.32
conspiracy	1.1	-4.24	3.7	-13.96	3.72
new	1.03	-5.13	5.4	-18.93	6.39
update	0.87	-0.15	2.5	-6.57	5.69
cases	0.83	-12.37	6.3	-26.75	2.10
outbreak	0.72	-1.25	2.9	-8.31	5.66

Table 3: Top eleven features from the random forest model with the conspiracy and title as feature with the statistics of the coefficients' posterior probability distributions. The first column shows the percentage of feature importance.

	svm	**mean**	**SD**	**1%**	**99%**
conspiracy %	2.82	**34.96**	6.2	20.56	50.09
virus	0.93	-6.70	5.3	-19.64	4.82
covid19	0.84	**-28.8**	10	-54.33	-6.20
god	0.75	**19.29**	7.6	3.39	37.54
allah	0.73	-40.09	26	-103.18	1.32
china	0.72	-4.64	3.9	-14.60	3.76
gates	0.69	3.39	16	-32.39	42.94
amir	0.68	-8.57	6.6	-24.66	5.81
lab	0.68	**-20.70**	8.2	-40.57	-2.28
cases	0.66	-22.41	14	-57.26	8.48
trump	0.63	14.53	9.6	-7.23	36.92

Table 4: Top eleven features from the SVM model with conspiracy and first 100 comments as features with the statistics of the coefficients' posterior probability distributions. The first column shows the SVM coefficients.

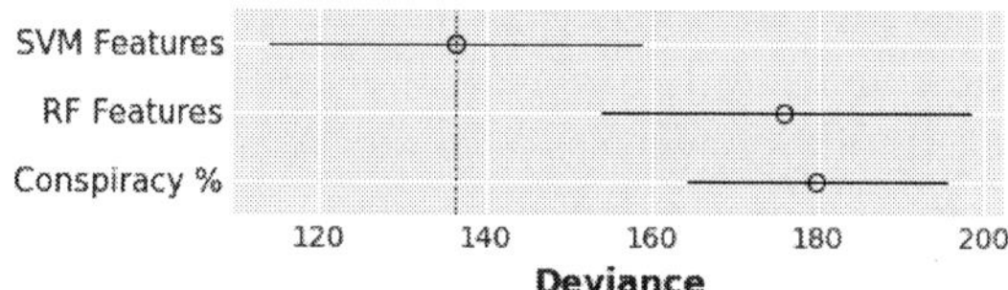

Figure 3: Deviance using WAIC as model selection metric. Black error bars represent the standard error.

content of the videos' first 100 comments significantly improves the accuracy of misinformation video detection from 82.2 to 89.4. However, in large-scale settings, it may be prohibitive to store the raw comments and continuously perform batch classification. In contrast, the conspiracy percentage feature only requires storing one conspiracy comment counter per video. Future research could leverage the video content to increase the classifier accuracy. The detection of misinformation on social media remains an open challenge, and further research is needed to understand how the COVID-19 misinfodemic spread to prevent future ones.

References

Jason Baumgartner, Savvas Zannettou, Brian Keegan, Megan Squire, and Jeremy Blackburn. 2020. The pushshift reddit dataset. In *Proceedings of the International AAAI Conference on Web and Social Media*, volume 14, pages 830–839.

Emily Chen, Kristina Lerman, and Emilio Ferrara. 2020. Covid-19: The first public coronavirus twitter dataset. *arXiv preprint arXiv:2003.07372*.

Sohan De Sarkar, Fan Yang, and Arjun Mukherjee. 2018. Attending sentences to detect satirical fake news. In *Proceedings of the 27th International Conference on Computational Linguistics*, pages 3371–3380, Santa Fe, New Mexico, USA. Association for Computational Linguistics.

Marco L Della Vedova, Eugenio Tacchini, Stefano Moret, Gabriele Ballarin, Massimo DiPierro, and Luca de Alfaro. 2018. Automatic online fake news detection combining content and social signals. In *2018 22nd Conference of Open Innovations Association (FRUCT)*, pages 272–279. IEEE.

Jacob Devlin, Ming-Wei Chang, Kenton Lee, and Kristina Toutanova. 2018. Bert: Pre-training of deep bidirectional transformers for language understanding. *arXiv preprint arXiv:1810.04805*.

A Seza Doğruöz, Natalia Ponomareva, Sertan Girgin, Reshu Jain, and Christoph Oehler. 2017. Text based user comments as a signal for automatic language identification of online videos. In *Proceedings of the 19th ACM International Conference on Multimodal Interaction*, pages 374–378.

Carsten Eickhoff, Wen Li, and Arjen P De Vries. 2013. Exploiting user comments for audio-visual content indexing and retrieval. In *European Conference on Information Retrieval*, pages 38–49. Springer.

Emilio Ferrara. 2020. What types of covid-19 conspiracies are populated by Twitter bots? *First Monday*.

Katja Filippova and Keith B Hall. 2011. Improved video categorization from text metadata and user comments. In *Proceedings of the 34th international ACM SIGIR conference on Research and development in Information Retrieval*, pages 835–842.

Sheera Frenkel, Ben Decker, and Davey Alba. 2020. How the 'plandemic' movie and its falsehoods spread widely online.

Tarleton Gillespie. 2018. *Custodians of the Internet: Platforms, content moderation, and the hidden decisions that shape social media*. Yale University Press.

Pankaj Gupta, Khushbu Saxena, Usama Yaseen, Thomas Runkler, and Hinrich Schütze. 2019. Neural architectures for fine-grained propaganda detection in news. In *Proceedings of the Second Workshop on Natural Language Processing for Internet Freedom: Censorship, Disinformation, and Propaganda*, pages 92–97, Hong Kong, China. Association for Computational Linguistics.

Andreas Hanselowski, Avinesh PVS, Benjamin Schiller, Felix Caspelherr, Debanjan Chaudhuri, Christian M. Meyer, and Iryna Gurevych. 2018. A retrospective analysis of the fake news challenge stance-detection task. In *Proceedings of the 27th International Conference on Computational Linguistics*, pages 1859–1874, Santa Fe, New Mexico, USA. Association for Computational Linguistics.

Chunneng Huang, Tianjun Fu, and Hsinchun Chen. 2010. Text-based video content classification for online video-sharing sites. *Journal of the American Society for Information Science and Technology*, 61(5):891–906.

Shan Jiang, Miriam Metzger, Andrew Flanagin, and Christo Wilson. 2020. Modeling and measuring expressed (dis) belief in (mis) information. In *Proceedings of the International AAAI Conference on Web and Social Media*, volume 14, pages 315–326.

Shan Jiang and Christo Wilson. 2018. Linguistic signals under misinformation and fact-checking: Evidence from user comments on social media. *Proceedings of the ACM on Human-Computer Interaction*, 2(CSCW):1–23.

Young Bin Kim, Jun Gi Kim, Wook Kim, Jae Ho Im, Tae Hyeong Kim, Shin Jin Kang, and Chang Hun Kim. 2016. Predicting fluctuations in cryptocurrency transactions based on user comments and replies. *PloS one*, 11(8):e0161197.

Sumeet Kumar and Kathleen Carley. 2019. Tree LSTMs with convolution units to predict stance and rumor veracity in social media conversations. In *Proceedings of the 57th Annual Meeting of the Association for Computational Linguistics*, pages 5047–5058, Florence, Italy. Association for Computational Linguistics.

Heidi Oi-Yee Li, Adrian Bailey, David Huynh, and James Chan. 2020. YouTube as a source of information on covid-19: a pandemic of misinformation? *BMJ Global Health*, 5(5).

Quanzhi Li, Qiong Zhang, Luo Si, and Yingchi Liu. 2019. Rumor detection on social media: Datasets, methods and opportunities. In *Proceedings of the Second Workshop on Natural Language Processing for Internet Freedom: Censorship, Disinformation, and Propaganda*, pages 66–75, Hong Kong, China. Association for Computational Linguistics.

Yinhan Liu, Myle Ott, Naman Goyal, Jingfei Du, Mandar Joshi, Danqi Chen, Omer Levy, Mike Lewis, Luke Zettlemoyer, and Veselin Stoyanov. 2019. Roberta: A robustly optimized bert pretraining approach. *arXiv preprint arXiv:1907.11692*.

Elaheh Momeni, Claire Cardie, and Myle Ott. 2013. Properties, prediction, and prevalence of useful user-generated comments for descriptive annotation of social media objects. In *Seventh International AAAI Conference on Weblogs and Social Media*.

Souneil Park, Minsam Ko, Jungwoo Kim, Ying Liu, and Junehwa Song. 2011. The politics of comments: predicting political orientation of news stories with commenters' sentiment patterns. In *Proceedings of the ACM 2011 conference on Computer supported cooperative work*, pages 113–122.

Verónica Pérez-Rosas, Bennett Kleinberg, Alexandra Lefevre, and Rada Mihalcea. 2018. Automatic detection of fake news. In *Proceedings of the 27th International Conference on Computational Linguistics*, pages 3391–3401, Santa Fe, New Mexico, USA. Association for Computational Linguistics.

Kashyap Popat, Subhabrata Mukherjee, Andrew Yates, and Gerhard Weikum. 2018. DeClarE: Debunking fake news and false claims using evidence-aware deep learning. In *Proceedings of the 2018 Conference on Empirical Methods in Natural Language Processing*, pages 22–32, Brussels, Belgium. Association for Computational Linguistics.

Vahed Qazvinian, Emily Rosengren, Dragomir Radev, and Qiaozhu Mei. 2011. Rumor has it: Identifying misinformation in microblogs. In *Proceedings of the 2011 Conference on Empirical Methods in Natural Language Processing*, pages 1589–1599.

Sarah T Roberts. 2019. *Behind the screen: Content moderation in the shadows of social media*. Yale University Press.

Victoria Rubin, Niall Conroy, Yimin Chen, and Sarah Cornwell. Fake news or truth? using satirical cues to detect potentially misleading news.

Kai Shu, Amy Sliva, Suhang Wang, Jiliang Tang, and Huan Liu. 2017. Fake news detection on social media: A data mining perspective. *ACM SIGKDD explorations newsletter*, 19(1):22–36.

B Silverman. 2019. Crowdtangle for academics and researchers.

Yla R Tausczik and James W Pennebaker. 2010. The psychological meaning of words: Liwc and computerized text analysis methods. *Journal of language and social psychology*, 29(1):24–54.

James Thorne and Andreas Vlachos. 2018. Automated fact checking: Task formulations, methods and future directions. In *Proceedings of the 27th International Conference on Computational Linguistics*, pages 3346–3359, Santa Fe, New Mexico, USA. Association for Computational Linguistics.

Aki Vehtari, Andrew Gelman, Daniel Simpson, Bob Carpenter, and Paul-Christian Bürkner. 2019. Rank-normalization, folding, and localization: An improved r for assessing convergence of mcmc. *arXiv preprint arXiv:1903.08008*.

Svitlana Volkova, Kyle Shaffer, Jin Yea Jang, and Nathan Hodas. 2017. Separating facts from fiction: Linguistic models to classify suspicious and trusted news posts on Twitter. In *Proceedings of the 55th Annual Meeting of the Association for Computational Linguistics (Volume 2: Short Papers)*, pages 647–653.

Sumio Watanabe and Manfred Opper. 2010. Asymptotic equivalence of Bayes cross validation and widely applicable information criterion in singular learning theory. *Journal of machine learning research*, 11(12).

Zhilin Yang, Zihang Dai, Yiming Yang, Jaime Carbonell, Russ R Salakhutdinov, and Quoc V Le. 2019. Xlnet: Generalized autoregressive pretraining for language understanding. In *Advances in neural information processing systems*, pages 5753–5763.

Xinyi Zhou and Reza Zafarani. 2018. Fake news: A survey of research, detection methods, and opportunities. *arXiv preprint arXiv:1812.00315*.

Association for Computational Linguistics
209 N. Eighth Street
Stroudsburg, Pennsylvania 18360

ISBN 978-1-7138-2048-2